LESSONS
FOR
SURVIVAL

LESSONS
FOR
SURVIVAL

MOTHERING AGAINST "THE APOCALYPSE"

EMILY RABOTEAU

HENRY HOLT AND COMPANY (H) NEW YORK

Henry Holt and Company
Publishers since 1866
120 Broadway
New York, New York 10271
www.henryholt.com

Henry Holt® and ⒽJ® are registered trademarks of Macmillan Publishing Group, LLC.

All photographs courtesy of the author unless otherwise indicated. Images on pages 255, 257, 259, 260, 262, and 266 are from the score of "Gut Bucket Blues," 1925, composed by Louis Armstrong, in Lil Hardin's hand, courtesy of the Library of Congress.

Portions of the book have appeared in the following publications in different form: *Aster(ix), Best American Science and Nature Writing 2021, Best American Travel Writing 2020, Bomb, BuzzFeed Reader, Gay Magazine, Goodbye to All That (Revised Edition): Writers on Loving and Leaving New York, Literary Hub,* the *Nation,* New York, the *New Yorker,* the *New York Review of Books,* the *New York Times, Orion,* and *The Fire This Time: A New Generation Speaks About Race.*

Library of Congress Cataloging-in-Publication Data

Names: Raboteau, Emily, author.
Title: Lessons for survival : mothering against "the apocalypse" / Emily Raboteau.
Other titles: Mothering against "the apocalypse"
Description: First edition. | New York : Henry Holt and Company, 2024.
Identifiers: LCCN 2023054865 | ISBN 9781250809766 (hardcover) | ISBN 9781250809773 (ebook)
Subjects: LCSH: Raboteau, Emily—Family. | African American women— New York (State)—New York—Biography. | African American mothers— New York (State)—New York—Biography. | African Americans—New York (State)—New York—Biography. | Climatic changes—New York (State)— New York—History—21st century. | New York (N.Y.)—Race relations— History—21st century. | New York (N.Y.)—Biography.
Classification: LCC F128.57.R34 A3 2024 | DDC 974.7/00496073092 [B]—dc23/ eng/20231205
LC record available at https://lccn.loc.gov/2023054865

Our books may be purchased in bulk for promotional, educational, or business use. Please contact your local bookseller or the Macmillan Corporate and Premium Sales Department at (800) 221-7945, extension 5442, or by e-mail at MacmillanSpecialMarkets@macmillan.com.

First Edition 2024

Designed by Meryl Sussman Levavi

Printed in the United States of America

1 3 5 7 9 10 8 6 4 2

To Geronimo, whom we named after the warrior,
and to Ben, who named himself,
I offer this quilt.

If we want to be good ancestors, we should show future generations how we coped with an age of great change and great crises.

—Dr. Jonas Salk

Live. Hold out. Survive. I don't know whether good times are coming back again. But I know that won't matter if we don't survive these times.

—Octavia Butler

CONTENTS

PART V. THE THING WITH FEATHERS

PREFACE

On my way through Times Square to my baby shower, I came across an ominous sign for the Rapture. Even among the thousand more bright and blinking ads in the noisy surround, this one popped out. It was written in chalk on a sidewalk sandwich board:

> **THE END**
> **OF THE**
> **WORLD**
> **IS NIGH!**
> **5/21/2011**

That was my baby's due date.

I laughed off the sign as evangelical mumbo jumbo. How many times before had the apocalypse failed to arrive? I didn't understand then that the world as I had known it was indeed about to end. Keeping an infant alive with the milk of my own body would check my personal freedom, exhaust my energy, and nearly break my psyche. I couldn't foresee the turbulent era about to unfurl, nor really, the high price of love. I remember feeling expectant when I descended the stairs to my friend Cathy's garden apartment on Forty-Third Street. I felt radiant and held.

The living room was full of the women in my circle. They were single mothers and mothers of adopted children and stepmothers and child-free women who had themselves been mothered and mothers of babies who'd never been born. They came from different economic classes, and so I'd asked my women to bring parenting wisdom instead of material gifts, unless the gifts were homemade.

Cathy's place was decorated with African cloth, hanging from the windows and upholstering the chairs. Sweet Honey in the Rock played from a speaker. A family portrait of the Obamas hung on the wall.

My writing partner, Miranda, had somehow managed to cook hors d'oeuvres despite having a toddler underfoot. My stepmother had made a mobile to hang above the crib. My auntie Marlene was there, but also not, because she had the beginnings of dementia. My mother-in-law, who was still parenting an adult child, said that I was going to need a lot of help. My colleague Drema said that I should throw my ideas of a picture-perfect family out the window. My college roommate Nicole said that I should put on my own oxygen mask first. "Fake it 'til you make it," she added. Cathy, who had recently divorced, said that I should never judge another woman's parenting because parenting was too goddamned hard. My soul sister, Angie, who'd survived a difficult childbirth, said to find the comadres. My high school friend Rachel, who'd decided not to have kids, said to hold on to my sense of self. My brother's wife, Cara, said to enjoy the early years, which would go both slow and fast at once. Finally, it was my mother's turn to speak.

"There's never a good time to have kids," she said.

We all laughed because that was the opposite of what a greeting card might say. I appreciated that my mother could admit to her own ambivalence and speak the truth. There are always challenges, internal and external, to mothering, no matter the time, history, or place. In my case, raising children has primarily spanned the second decade of this century, a turbulent time that included the Trump presidency, social division, civil strife, culture wars, stricter immigration policies, the separation of families at the border, the Black Lives Matter movement, the Me Too movement, the dismantling of reproductive rights, the swiftening of the Great Acceleration, worldwide awakening to the climate crisis, the global pandemic and, in my family, the loss of my father, who was also my muse.

This unsettling decade coincided, for me, with what they call the "sandwich years," in which caring for young children and aging parents at the same time can lead to burnout, isolation, and guilt. Truly, this was not a good time to have kids. What does it mean to occupy the complex position of "mother," principally responsible for bringing life into a fallen world? From a place of disequilibrium and periodic depression, in the context of multiple threats and interrelating crises, like many parents—like my own parents, and their parents before them—I struggled. I'm struggling still, to raise my children to thrive without coming undone.

There is no linear narrative for what we're living through, another friend recently said. How to make meaning by reading the signs of the times? In the United States of America, our narrative of progress is in deep crisis, connected to the failure of society and culture to make sense of our history of genocide and slavery, which have led to ecocide and biodiversity loss. This book is unified by my search for "lessons for survival" in dark times, my devotion to exploring, and my ongoing interests in the meaning of home, the relationship of the inner city to the city, the common good, public space, public art, small acts of witness, social and environmental justice, and the radical potential of mothering.

What does it mean to survive in the midst of protracted crises; to continually renegotiate threats against life; to cope; to stare grim asymmetries of power in the face; to connect with struggles beyond this country that too often endangers Black life; to build community; to find shelter; to apply privilege toward engagement with civics, commitment to the common good, and bringing up children to be good citizens? All of these interrogations are driven by the question that keeps me up at night: *Will my children be all right when I'm gone?*

At my baby shower, my mother went on to recall the day I was born, how much meaning I'd brought to her life, and to reflect on how much more meaning I was bringing her now in becoming a mother myself. Then she gave me the gift she'd sewn for her grandchild—a patchwork Log Cabin quilt. After the birth of my second child, under two years later, she sewed another, using some of the same cloth as the first but in a different pattern, her own improvisation on Flying Geese.

This book is structured something like a quilt, pieced together out of love by a parent who wants her children to inherit a world where life is sustainable. I must thank my mother here, at the outset of the book, for minding my children so often while I journeyed—both far afield and closer to home—to seek its constituent parts.

—EMILY RABOTEAU
2023
The Bronx

LESSONS
FOR
SURVIVAL

PART I
A BIRD'S-EYE VIEW

American Redstart, 3612 Broadway, Sugar Hill, Harlem,
muralist: James Alicea

Burrowing Owl, 606 W. 145th St., Sugar Hill, Harlem,
muralist: Jana Liptak

SPARK BIRD

If I can be called a bird-watcher, my spark was a pair of burrowing owls, painted on the narrow storefront gate of a shuttered real estate business on 145th Street, in Harlem, that brokers single-room-occupancy housing for two hundred dollars a week. I spotted them after ice-skating with one of my kids, at the rink in the shadow of towering smokestacks at Riverbank State Park, a concession to our community for the massive wastewater treatment plant hidden beneath it. It was midway through the Trump years: January, but not cold like Januaries when I was little, not cold enough to see your breath. It wasn't snowing, and it wasn't going to snow. The owls watched me quizzically, with their heads cocked, their long, skinny legs perched on the colored bands of a psychedelic rainbow that seemed to lead off that gray street into another, more magical realm.

Among people who watch birds, it's often the case that a first bird love, the so-called "spark bird," draws them forever down the bright and rambling path of birding. For Aimee, it was the peacocks in her grandmother's backyard in southern India. For Kerri, it was a whooper swan above Inch Island, in County Donegal, the year the peace process began. For Windhorse, it was the Baltimore orioles flitting about in the high branches of poplars at his grandfather's house up north, on the lake. For Meera, it was the red-winged blackbird, there at the feeder, when she was small. Her mom told her the name, and it all clicked into place—*black bird, red wings*—as she learned the game of language and how we match it to the world around us.

I pointed out the extraordinary owls to my kid, stopping to take a picture with the camera on my phone.

"Look," I said.

"I want hot cocoa," my kid replied.

We turned the corner and made our way up Broadway toward the Chipped Cup for overpriced Belgian hot chocolate. On the corner of 149th, I spotted another bird, the American redstart. It was painted on the security gate of Washington Heights Pediatrics—the kind of doctor's office that struggles to keep the lights on with trifling Medicaid payments and nebulizes asthmatic Black and brown kids, like mine, with albuterol when their lungs constrict too severely for a pump to clear at home. The tuck of color under its wing matched my kid's unnecessary winter hat. I took another picture. Oh, New York—you gorgeous aviary of madcap design! Across the street, from the corner of my eye, I saw more: a pair of Calliope hummingbirds painted mid-flight outside the Apollo Pharmacy.

That's when I understood there was a pattern.

Calliope Hummingbird, 3659 Broadway, Sugar Hill, Harlem, muralist: Kristy McCarthy

After the spark, I started noticing scores of them along my two-mile walk to work at City College. Most of the bird murals in Upper Manhattan are spray-painted on the rolled-down security gates of mom-and-pop shops along the gallery of Broadway, at street level. Others are painted up higher, on the sides of six-story apartment buildings. They nest, perch, and roost in the doorways of delis, pharmacies, and barbershops. Lewis's woodpecker at the taqueria, the almighty boat-tailed grackle at the

Buena Vista Vision Center, Brewer's blackbird at La Estrella Dry Cleaners, and so on—there are dozens of bird murals, each one marked in a corner with the name of the ongoing series to which they belong: the Audubon Mural Project.

John James Audubon, the pioneering ornithologist and bird artist, once lived in the hood. He's buried in the cemetery of Trinity Church, at 155th Street, midway between my apartment building in Washington Heights and my job in Sugar Hill, Harlem, where I teach writing, sometimes using Wallace Stevens's "Thirteen Ways of Looking at a Blackbird" as a prompt. Audubon Terrace, once part of his estate, is now the site of a complex of cultural buildings. Other uptown locales named after Audubon include a housing project, an avenue, and the ballroom where Malcolm X was assassinated. Its historic facade remains as cladding to a newer medical research building—a concession to Black Americans who protested the ballroom's demolition—at 165th Street, across from the emergency room of a New York–Presbyterian hospital. When you walk by these places, as I do, you can spy many of the same birds Audubon chronicled in his masterful archetype of wildlife illustration, *Birds of America* (1827–1838), in the guise of public art.

The project is an unfolding environmental awareness partnership among the gallerist Avi Gitler, the National Audubon Society, and local business and property owners. The murals are sponsored through donations to Audubon and painted by myriad artists, some of them local, in a diverse range of styles. Uptown, there are presently 123, and counting, bird species depicted. (Sometimes they disappear when businesses change hands.) The project aims to reach 389. This is the number of North American species, according to Audubon's 2019 "Survival by Degrees" report on birds and climate, at risk of extinction from climate change—an alarming two-thirds of North American birds. I have attempted, so far, to photograph them all. The world is changing faster than we can. The changes are restructuring our lives in ways we struggle to respond to, birds and humans alike. How do we navigate this shifting terrain?

A printable map on the Audubon Society's website indicates the address of each mural. I prefer not to use that resource as a guide; I like the element of surprise. As with actual birders, I never know which birds

I'll see on a walk. Sometimes a new bird appears to have landed over-night. Older birds may be marked with graffiti or sullied by weather and grime. I was saddened to discover from the window of the M4 bus, while riding downtown, that someone had spray-painted over the tundra swan I'd come to love with a cloudy white cipher of bubble letters. *Who did that?* I wondered, thinking of that rogue graffiti artist known as Spit in the 1984 hip-hop movie *Beat Street*, who defaced the work of other art-ists by tagging over it.

I felt glad to have documented the tundra swan before it disappeared. If temperatures rise three degrees Celsius, 93 percent of this bird's breed-ing habitat in the tundra of far northern Canada is projected to be lost. Because the Arctic is warming faster than anywhere else on the planet, tundra swans have nowhere farther north to go.

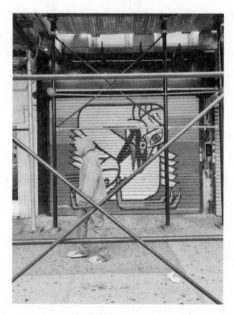

Tundra Swan, 3631 Broadway, Washington Heights, Manhattan,
muralist: Boy Kong

As a photographer, I am drawn to the visual echoes between the fash-ion and the feathers, the postures of people and wildlife: for example, the sweep of a dark trench coat that seems to give motion to the pere-grine falcon's wings, or the pair of black track pants that have merged with the legs of the glossy ibis so that the young man wearing them appears to be riding the bird. I wish to document the tensions among human, bird, art, and commercial signage. Sometimes the artists play with these elements, too, as with Snoeman's mural of a Canada goose, wherein the bird beneath the canopy of a shoe store is styled in a pair of Timberland boots.

Canada Goose, 3868 Broadway, Washington Heights, Manhattan,
muralist: Snoeman

I understand the project has landed in this neighborhood because of its connection to Audubon, and also because of the millions of endangered birds that migrate above Manhattan and continue to nest within it. I also appreciate its potential for helping connect a low-income community of color to the green sector, which is predominantly white.

Amelia Earhart is quoted in the mural at Manuel's Grocery, at 152nd Street: "No borders, just horizons. Only freedom." The bright yellow breast of the mangrove cuckoo pictured there matches the tank of the blowtorch in the hand of the plumber passing by. That bird is described as "a rare bird native to the Dominican Republic, Puerto Rico, and Florida in the U.S." Unlike nations preoccupied with immigration, the artist states, "Birds See No Borders."

I love these birds for their beauty, the way birders love actual birds, for the exalted brushstrokes of their wingspans that lift us from the drudge of survival. Some birds are reclusive: for example, the Florida scrub-jay and Mexican jay have long been trapped behind a hunter-green construction fence. The birds on buildings under scaffolding look caged. At businesses that struggle to pay escalating rents by staying open for twelve hours a day, seven days a week, the birds can be seen only at night, when the gates come down. At shops that have closed and not yet reopened, like the beauty salon with the laughing gull, the bird is always there.

Laughing Gull, 1728 Amsterdam Ave., Sugar Hill, Harlem, muralist: Simon Aredondo

"We know that the fate of birds and people are intertwined," *Audubon* magazine editor Jennifer Bogo, wrote me. "That's especially true in communities, like northern Manhattan, that suffer disproportionately from environmental and human health burdens. We hope that the Audubon Mural Project makes people literally stop in the streets and consider what's at stake with this critically important planetary crisis, while at the same time beautifying and drawing attention to neighborhoods that have historically not been the focus of environmental protections."

To my eye, the project is at once a meditation on impermanence, seeing, climate change, environmental justice, habitat loss and a sly commentary on gentrification, as many of the working-class passersby are being pushed out of the hood, in a migratory pattern that signals endangerment. Most of all, the murals bring me wonder and delight. I can hardly be called a bird-watcher. But because this flock has landed where I live, work, parent, pray, vote, and play, permit me to be your guide.

Peregrine Falcon, 752 St. Nicholas Ave., Sugar Hill, Harlem, muralist: Damien Mitchell

Glossy Ibis, 3671 Broadway, Washington Heights, Manhattan, muralists: Kristy McCarthy and Pelumi Adegawa

2021

Caution, N. Moore St. and Hudson St., TriBeCa, Manhattan,
artist: Dee Dee Was Here

CAUTION

I needed the birds because I was in pain. When did I start feeling the pain? I think it was when our republic elected the demagogue and hope went missing. I remember the presidential debate: she, in her white pant-suit; he, pacing behind her like a menacing ape. The familiarity of that dance. *He's going to hit her*, I thought. I clenched my shoulders. In the years I am talking about, 2016–2020, they remained clenched.

"My God," my husband marveled when the election was called. "America really hates women." I had told our two kids at bedtime that in the morning they'd wake to our country's first female president. "You lied to us!" the five-year-old cried at breakfast, betrayed.

The election inspired a feeling of heartbreak bordering on insanity, which I dared not reveal to my family lest my family lose hope, too. For the most part, I carried the grief alone. On occasion, I vented to girl-friends, who got it. Angie gave me her copy of *The Body Keeps the Score*. Our autonomic nervous systems were all riled up. We were inflamed. The external calamity, which was political, coincided with the internal calamity, which was personal. My country, my body: broken. *Me, too,* the women said.

I looked all over the city for relief, not yet noticing the birds. My journey began with a chiropractor on the Upper West Side. Before he cracked my neck, the chiropractor told me I had an atlas tilt at the top of my cervical spine: the weight of the world on my shoulders. It was unclear to me whether the atlas could be manipulated back into place on its axis.

Ori Mi Pe, the classic Adire cloth indigo design, literally translates as "my head is correct." In Yoruba belief, the head is the seat of personal destiny. Therefore, Ori Mi Pe means, "I will have a good destiny." But my head was not correct. The junction of my skull and spine was off, the nerve center misfiring in that cathedral of delicate bones. I wondered what this meant about my destiny.

"Cervix" refers to the neck, or any necklike part, especially the constricted lower end of the uterus.

When he was deep inside me, my lover, my predator, claimed that he could feel my cervix. What did it matter that the abuse was years in the past, well beyond any statute of limitations—back when a hit song on the radio was "Survivor," by Destiny's Child; back in the days when I sat in a Survivors Anonymous peer support group of other women with PTSD in a church basement in Brooklyn—when it felt like the violence was happening all over again, right now?

"Do you feel guilty?" asked the primary care physician before referring me to a psychotherapist. I was already seeing a therapist on One Hundredth Street and Central Park West, who'd insisted on blood work and recommended bringing my husband to vouch for the truth of my pain, since doctors, as a general rule, disbelieve women. My husband was busy. I was busier than him by far, and on top of that, I was sick. The blood test could not explain why my hair was falling out. Thyroid, endocrine system, hormones: all were in normal range. And yet, I could barely get out of bed.

I wasn't free to stay in bed. Duty called. My children needed food and feeding, washing, cuddles, stories, protection from the incessant evil messaging of white folks, and lotion on their ashy knees. My students needed feedback, mentoring, encouragement, recommendation letters, and advice on handling the sexual advances of their male professors. My health insurance needed explanations. My spine was either the sum of my moods, a barometer of the era, or a vertical timeline of historical abuse.

The orthopedic surgeon barely glanced at the X-rays before accusing me of being a boring patient. "You must have a lot of time on your hands," he said, dismissing my symptoms, "to be worrying over nothing."

The pain, he implied, was all in my head.

I did not have a lot of time on my hands. I had next to no time to myself, practically none at all. But it felt as though I'd been in a car crash. *I am the backbone of my family*, I wanted to tell him, *how dare you speak to me like that; I am the backbone of my community. I birthed two babies at home without drugs because I trusted my own body to be a mammal more than I trusted in a healthy outcome from the medical machine. I am an educated woman suffering private heartache under a dictatorship.*

These trips to doctors were fitted between trips with the children to the playground; to the Bronx Zoo; to Coney Island; to their barber, José; to the American Museum of Natural History; to the New York Aquarium; to the public library; to church; to their grandmother's house in Queens; to the pediatrician; to the apple orchard upstate; back to the public library; back to the playground; and to the Little Red Lighthouse beneath the George Washington Bridge.

The symptoms radiating down my nervous system like falling dominoes included brain fog, tension headaches, jaw pain, limited head rotation, stiff neck, shoulder pain, tennis elbow, carpal tunnel syndrome, pelvic asymmetry, asthma, protrusion of discs, chronically cold hands and feet, low blood pressure, low self-confidence, paranoia, chronic diarrhea, despair, and fatigue. Worst of all, I could no longer hold up my head.

The dentist said I was like the princess and the pea. My teeth conveyed no rationale for the throbbing pain in my jaw. He accused my mouth of having perfect teeth but nevertheless referred me to an oral surgeon with a bad bedside manner. I told him I couldn't sleep. We were on Eighty-Sixth Street, near the site of Seneca Village, where two hundred years ago a thriving Black enclave was displaced by the city through eminent domain to make Central Park.

You're like the princess and the pea, the oral surgeon said, with the same disdain as the dentist. The two of them were in a private club; they shared a language, a philosophy predicated on the put-down. Neither man would prescribe painkillers. In spite of my straight, white teeth, they suspected that I was an addict hustling for opiates, that my mouth was a den of lies. I didn't want painkillers. I wanted the pain to end.

The pain traveled. While I searched up and down the island of Manhattan for care, countless women, including my mother, gathered in the cities, and in the capital, and across the globe, wearing pink pussy hats. We seemed to intuit already the overturning of *Roe* and the impact that would have on our lives. Meanwhile, my mostly supportive husband asked me to leave our apartment so that he could play violent video games in solitude. *Call of Duty*. He was granted an award for being a genius. When the school complained about the behavior of our son, it was me they called, never him. Meanwhile, it seemed to me that the ripped subway posters I encountered on my multiple trips to doctors as I crisscrossed the city were speaking to me in code. Sometimes they appeared as abstract or conceptual art in the style of nouveau réalisme, as good or better than anything you'd find at the Guggenheim, except that the artist was unknown; the artist was many people, the exhausted of the earth, bored as they waited for the train, tearing away at the advertising; the artist was the grit of the underground, the brashness of the city, and time itself performing décollage.

Empire State of Mind: A Raymond Hains Rip-Off,
135th St., Harlem, downtown C train platform

I was not alone in my illness, nor alone in being blamed for my illness. The illness was chronic. We grew fibroids. Nobody knew why more Black and brown women were afflicted at a higher rate. From Audre Lorde: "When you live on the edge of any structure, you have to know that survival is not theoretical." Our uterine linings began appearing in unexpected parts of our bodies, including our brains. Endometriosis. Prolapse. Fibromyalgia. Some of us dissociated. Some of us had panic attacks. The level of cortisol in our bloodstreams grew toxic. They removed the uterus of my friend. They put children in cages at the border. We were angry at the lies of women's lib and civil rights, at the failed experiment of our country. We admitted to ourselves that white people could not be redeemed. We were taunted by our president's tweets. Unsure what to do with our rage, we turned it on ourselves.

Built for Glory,
116th St., Harlem, downtown C train platform

Our immune systems attacked us. We grew tumors. *Me, too*, we repeated. *Me, too.* The previous season we'd begun saying, *Black lives matter.* Nobody listened. Nobody knew how to balance our hormones. Probably there was not enough money in it. They theorized it was because we were not bearing enough children. We begged for estrogen and were refused. Our perception was either extremely distorted or crystal clear. Our trauma was too complex to diagnose.

Our cells multiplied, rampantly, into more tumors. They tried to remove the one remaining breast of my friend, who insisted on keeping it for the sake of her sex appeal, for the sake of her self and her love of sex, and somehow I understood her warrior stance was rooted in her being Ethiopian. As for the rest of us, unless we were queer, we were having disappointing sex, or torrid, self-destructive love affairs, or no sex at all.

When my senior colleague asked me to wear the pair of fancy lace panties he'd brought back from Paris at the next faculty meeting, I could have reported him. Maybe I should have. But I didn't trust the system to take him down. Instead, I kept the panties and bided my time. They were as multicolored as a peacock feather. They were worthy of Josephine Baker. They were the sexiest panties I've ever owned. Wearing them was a misdemeanor of survival.

During the hearing of the scoundrel accused of sexual assault who was nominated to the Supreme Court, I was waiting at the hospital in radiology for my turn in the MRI machine. CNN played in the waiting room. Everyone present was in pain and disgust with our president. America didn't listen to the woman. The technician gifted me a pair of earplugs, but it was as loud as a uterus inside that white sarcophagus. The clacking noises sounded nearly human. "Dewey," repeated the voice, nonsensically. "God." "Dot." "Did I?" "I did."

I was not to move.

This was my second time in an MRI machine. The first was to examine a cystic tumor in my throat, years before. It was blocked anger, unreleased chi with nowhere to go, a homeopath told me. A surgeon cut it out of me, and now I wear a scar on my neck that makes me look tough, or victimized, depending on one's point of view. Back then, I hoped the magnetic field and radio waves generating images of my insides might be rearranging my atoms, that I might come out a different person—preferably a stronger one. But I did not.

Now, shoved back into the machine, I considered patterns of predation: A cervix. A throat. How much could be forced into those openings. How much shit could be swallowed. How could a body take it? I recalled my lover, my abuser, once telling me not to move. I would feel more pleasure if I kept still, he said before giving me an STD. "Dewey." "God." "Dot." "Did I?" "I did." Some people are triggered by the noise, they'd warned me outside the machine, some people get claustrophobic in there. *Push this button if you get scared.*

It came as no surprise when I was told that I was among the least qualified of the hundreds of applicants who competed for my position, and that I filled a demographic niche as a Black woman—that I should be grateful. I was the gift of diversity. I was living my ancestors' wildest dreams. Therefore, I should not behave as a diva.

The midwife who caught my children tried to teach me a new technique called "havening." She had me trigger the trauma, massage my own face with the tips of my fingers, and, while hugging myself, do the same to the tops of my arms, replicating a mother's touch; to mother myself. I was to repeat this mantra: *I deserve to feel safe.* I was to be my own shelter. I wanted it to work but admitted it did not. We were in Gowanus, near the toxic canal. "The trigger is too large," reasoned the midwife, "the trigger is patriarchy, and all the babies being born to women in this era will inherit the fight-or-flight response through the umbilical cords of their mothers. *Wait and see.*"

When my colleague became my boss, I reminded him of the panties. "Ever since you gave me that gift," I lied, "I've understood I could count on you as a friend." I asked him for something. I don't remember what. Time to myself? An office with a window? Respect? Whatever it was, I didn't get it.

The integrative health doctor advised me to quit running. We were in Chelsea, near the warehouse where they stored uranium for the Manhattan Project. "You're running with cement blocks on your feet," she said. "You have next to no bandwidth. You're like a vessel nearly empty of water. Your nervous system is trying to answer a question. You're not strong enough to run." After mixing these metaphors, she gave me the card of a somatic therapist who would teach me to refill my cup. I recognized the name on the card as belonging to a man who screwed my friend when she sought his help to get her head correct. I asked to see a woman and stopped training for races.

The somatic therapist told me she hadn't at first understood I was Black, pointing to a disconnect between how others sometimes see me outside the context of my Black family and who I understand myself to be. But now that she knew my mixed-race background, she said it was epigenetics causing my pain: the inherited trauma of my enslaved ancestors, the burdens of my grandmother, etc. . . . According to the latest studies, these injustices were encoded within me. We were down by Wall Street, near the African Burial Ground. She handed me a foam sledgehammer and encouraged me to use it to externalize my rage. This exercise embarrassed me, but I did not wish to hurt the well-meaning white woman's feelings. Performing my rage for her with that silly prop, I felt secretly enraged.

The burly physiatrist was so aggressively rude to his cowering receptionist that I walked out of his practice in Washington Heights before he could look at my body. I presumed that at home he beat his wife. The gold chain nestled in the black chest hair in the V-neck of his scrubs made me sick.

Patterns of predation were being amplified in the sloppy court of public opinion. The hunt for a cure was growing expensive. The diagnosis was inexact. My freedom of movement, that is, my range of motion, was restricted. This much I knew: I could not afford a broken back. Too many people depended upon me.

The osteopath was a good witch, I could tell. We were on Thirty-Second Street, not far from the Empire State Building. Her fingers felt nice at the base of my skull, and I did not exactly think her racist when she theorized that so many Black women in America are overweight because inherited trauma has made us warehouse our fat for the lean times. But by the time I left her padded table for the Korean market next to her practice to buy a pack of frozen black bean buns, my back was hurting again; the weight of it all was too much to bear; her juju was gone.

"You are not symmetrical," the acupuncturist said. We were in the West Fifties, a block from the Ed Sullivan Theater, where *The Late Show* is made. "Your muscles are calcified already. Your spine is no good. Your neck is no good. Your psoas is weak." He said he could treat the pain, 90 percent of which was in my head. I believed him 90 percent. Two times a week he needled me with insults. I pondered my pain and tried to breathe. That I could not relax, nor release, was my fault; that my muscles were in constant spasm, that my back had become a carapace, a shield, was my fault. The acupuncturist promised, quite arrogantly, to make the pain go away. On the wall hung a caricature of him with a frightened-looking porcupine who asks, "How many needles?"

A log of misalignment from those days, when I felt besieged:

> as I lie upon the acupuncture table
> as I wait for the train
> as I bathe with lavender soap
> as I fall asleep
> as I plank
> as I prolong getting out of bed in the morning
> as I lay out the children's school clothes
> as I pick lice from their hair
> as I visit my dad in the hospital
> as I peel the green apples
> as I arrange another playdate at the playground
> as I move the wet clothes from the washer to the dryer
> as the abuser is confirmed to high office
> as I breathe

as I attempt to write the novel
as I wait for the children's school bus
as I fold the laundry into separate piles

The highest spinal center is in the medulla oblongata, at the base of the brain, just above the place where the neck connects with the skull—the sixth chakra, through which cosmic energy nourishes the body with prana, conscious cosmic energy. I was the only Black woman in the yoga studio where the instructor spoke of chakras as the seven gates of freedom along the astral spine, after asking us to chant *om*, and I didn't forget it, that I was the only Black woman, for the entire class—not even in tree pose, where I balanced as straight as I could with a misaligned spine, nor in Savasana, where I lay on the floor like a corpse.

Death by a million microaggressions. By racism. By air pollution and the urban heat island effect. By disillusionment and the disintegration of kinship structures. By the police. By the man. By the world of men. By the weight of my grandmother's sorrow and the threat of the rising sea. By the needs of my needy children. By America. Death by the wolf of debt. A slow, inglorious death.

I Have Survived Every Attempt to Kill Me,
103rd St., Manhattan Valley, Upper West Side, uptown C train platform

Our body's question: What can we do when we can't just stop living? Down in TriBeCa I was taken by a work of street art by Dee Dee Was Here depicting a dreamy-eyed woman with a mouth muzzled by caution tape. The woman reminded me of the myth of Philomela, whose rapist cut out her tongue to stop her from denouncing him, and whom the gods transformed into a nightingale. She reminded me, too, of Escrava Anastacia, an enslaved woman whose masters forced her to wear an iron face mask but is now venerated as a folk saint in Brazil. Photographing her was my prayer for intercession in a time of trauma, an impulse to speak truth to power, and like the nightingale's song: a lament.

This is the last thing I am going to share about our pain, and I tell it only so that we who see ourselves as a superwoman, or a victim, or a problem, or merely as somebody's mother, and you who don't even see us at all, don't forget we have a body. There are other viewpoints on the city than a log of doctors who did not help. There are birds to find hiding in plain sight. And pilgrimages are most successful when they have a metaphysical purpose.

In the bestselling book on mind-body connection and healing back pain without drugs, surgery, or exercise gifted to me by my physical therapist, the author advises spending fifteen minutes every day reminding the brain that there is no structural problem, in a meditative state meant to induce myofascial release. Knowing that repressed anger and anxiety are the triggers, I was supposed to tell myself my back is fine. And yet,

the structural problem is real.

2020

PS 173 playground, Washington Heights, Manhattan

PLAYGROUND STRUCTURES

There are countless ways to see in a city of more than eight million. When my boyfriend left me with a bruised eye, I kicked him out, got a watchdog and a dog's-eye view of New York. I lived in Flatbush, Brooklyn, then. The dog was a stray, a mutt, the best kind of dog, half German shepherd and half something else that made his legs so short he appeared to be sinking in quicksand. He needed walking, so I walked him. I woke early to do it, or what I thought was early at the time. A sacrifice, but worth it for the exercise, the filter of the morning light, the conversations with strangers; worth it for the dew on the grass of Prospect Park when we, the dog owners, were permitted to let our pets off leash; worth it for the uncomplicated joy of the animals chasing each other, rolling in rotting leaves, chomping at snow, smelling each other's butts.

The park wasn't just a relief from the crush of the city but the city's great equalizer. New York's monomaniac master builder Robert Moses said of city planning, "As long as you're on the side of parks, you're on the side of the angels." In the park my mutt played on equal turf with highbred wolfhounds and common pit bulls. All of their shit stank the same, and all were uniformly loved. Most of the dogs were pampered. A few wore booties and matching jackets. I tied a red bandanna around my dog's scruffy neck. The kids on the block called him a Blood. I liked that he showed his teeth when men came over. He slept in my bed. For a time, we were a couple.

In those days, I worked as a secretary in an Episcopal church in Gramercy Park. The private two-acre oasis is only accessible to the rich residents of the buildings that border it. In the *New York Times*, journalist Charlotte Devree called Gramercy Park "a Victorian gentleman who has refused to die." Just 383 keys to the park exist. The church where I worked held one of them. On Fridays, my job was interrupted by loud protests as a group marched around the square's perimeter wielding handmade signs. "Parks are for the people," they yelled. On other days, pedestrians searched haplessly for the gate's entrance. Maybe from the key-holder's-eye view, the park looked most paradisiacal when the rabble was cast out. I dreamed of duplicating the key and distributing it like Halloween candy, but the key could not be copied. I never entered Gramercy Park. I had no truck with places that barred the commoners.

I moved to Harlem for a different job at City College, once known as the poor man's Harvard. Because the new building didn't allow pets, I didn't bring my dog. I returned him to the no-kill shelter and crossed my fingers that another New Yorker would take him in. So much for my principles. My address mattered to me more. From the rooftop, I could see Yankee Stadium twinkling like a jewelry box in the Bronx and all the zigzagging fire escapes like so many scars. Then I met a good guy.

Falling in love happened like this. We went on five or ten first dates, unsure if we were more than friends. On the last first date he asked what was up with the ratty red ribbon around my wrist. I'd gotten it in Salvador do Bahia, on the steps of the Igreja de Nosso Senhor do Bonfim years before, where a beggar woman selling prayer cards of Escrava Anastacia had grabbed my hand and knotted the ribbon at my pulse point three times, instructing me to make a wish for each knot. I was not to remove the ribbon if I wanted the wishes to come true, but rather to be patient, and let the bracelet fall off in time. I believed in magic enough to follow these instructions, but also, I confessed to my date, that by this point in my life's story, I was sick of the ribbon which had become a dirty string cramping my style.

"What did you wish for?" the man I would marry wanted to know. "I can't tell you that," I said, "but I can tell you this. One wish was for love, one was for family, and one was for work." In a smooth gesture, he took my hand, picked up his butter knife, cut off the ribbon and slipped it into his breast pocket. "I promise you," he said, patting his pocket, "that all your wishes will come true." I was astonished by his boldness. Did I want to kiss him or slap him? Hard to say. I decided to kiss him.

We moved into a co-op on 180th Street in Washington Heights, one block from the George Washington Bridge. We married. In short order I popped out two kids and got a parent's-eye view of New York.

Nowadays, I spend my time in playgrounds.

≡

The playgrounds are everywhere in this city, thank God, or else where would the kids play? But I hadn't seen them before. Or if I had, they were on my periphery—some fenced-off blacktop in a bigger green park, loud and packed with brats in the day, seedy and dark with shadows at night. There are a half dozen playgrounds within walking distance of our building, each with its own character based on the socioeconomics of the sub-neighborhood, but none equipped with the seesaws, merry-go-rounds, or tetherballs of my youth, now considered hazardous. My mother pities her grandkids. She thinks they require a yard attached to a house, like the one where I grew up in suburban New Jersey. But my kids spend more time outside than I ever did.

Dolphin Park, on our corner, is so small it doesn't rank on a map, not even as a freckle. It's open only in the summer for the baby and toddler set, with a sandbox full of plastic pails and shovels, a little playhouse, a short slide, and a dolphin statue that spits water from its rostrum. Quisqueya Playground, near the Fort Washington branch of the public library, hangs like a balcony over Harlem River Drive. It's got a swing set, wind chimes, feral cats, empty dime bags, shaved-ice pushcarts, and women hawking pastelitos or bootleg DVDs. Jacob Javits Playground lies at the foot of Fort Tryon Park. It has a peeling rainbow mural, balance beam, basketball courts, sprinklers, and greased-up musclemen doing chin-ups on the monkey bars. And that's just a partial list of the spots I can push a stroller to in under ten minutes.

In the playgrounds with my kids (as in Prospect Park with my dog), I talk with people I would otherwise never have spoken with: A Hasidic mother of six, a decade younger than I. A teenage mom, a decade younger than her. A Trinidadian nanny with a talent for sudoku puzzles. An out-of-work opera singer, father to twins. A foster parent peddling the *Watchtower*. Back when I owned the dog, I relished the public parks for their mixture of high and low. But now that I'm a parent, I despair of the divisions the playgrounds make plain. Having children in common, I realize, is not automatic grounds for friendship. Especially when the talk turns to schools, the inequality begins to penetrate.

"You're out of the good zone by one block," another mother tells me on a bench in Bennett Park. Bennett, in the "good" school zone, is the highest point of elevation in the city, at 265 feet above sea level. It boasts a two-million-dollar renovation, a spiraling slide, hanging rings, a children's garden, a hand-painted Little Free Library, and a replica of a cannon from the Revolutionary War, in memory of the base of operations General George Washington established at the summit to keep the enemy out.

This mom is a resident of the ritzy Castle Village apartment complex, with its river views, doormen, and twenty-four-hour security patrols. Her children attend nearby PS 187, in the area the realtors advertise as Hudson Heights. She suggests that I beg the principal of this elementary school for an exception to admit our son. The school he's zoned for, PS 173, is failing, she warns: substandard test scores, next to no parental involvement, an ineffective principal, overstuffed classes. She raises her well-manicured eyebrows and adds: "Ninety-nine percent Hispanic." There's a proposal to de-zone the district, but the parents on the local listserv are up in arms.

What listserv? Which district?

Our son is only three months old. I am sleep-deprived and only just getting the knack of breastfeeding him without my nipples shredding. I haven't yet studied the thirty-two zones of the Department of Education's map, which look an awful lot like gerrymandering, nor discovered that our school district is the most populous in the city, encompassing both Castle Village, on the western promontory over the river, and the Bridge Apartments housing project, whose ugly towers straddle the Trans-Manhattan Expressway, with the noxious fumes of the traffic below rendering the balconies unusable. I've yet to discover that this city has one of the most racially segregated and economically stratified school systems in the country. I merely want my boy to learn to crawl, walk, run, talk, and climb up to the slide over there that spirals like a single strand of the hair he'll eventually sprout.

I'd like to slap this woman, my neighbor, for her entitlement and presumption. Instead, I remark that long before she and I came to squat on this turf holding sippy cups, the Lenape got robbed of this island by the Dutch, and Washington lost this ground to the redcoats, and eventually the Irish, who settled the hood, were replaced by German Jews, who got scared off by the Blacks, who were then outnumbered by the Dominicans . . . By which I mean: nobody owns bupkis, and we're all just passing through. She and I don't become friends.

But later, when our son is ten months old and waking at the ungodly hour of 5:00 a.m., I drag him back to the playground in Bennett to wear him out. There, I meet another woman, a native New Yorker, with a son the same age. Our boys share raisins and bounce together on the squeaky bridge, testing out their legs. When I ask this woman what she hates most about motherhood, she laughs from her diaphragm. A man hollers down at us from his open window to shut the fuck up. We have woken him too early. The playground doesn't open officially until eight. "Come down here and make us, cocksucker!" my new friend shouts back.

One night, before we had children, my husband and I snorted cocaine with a swizzle stick from Starbucks. The drug seemed not to be working, and we talked for a few hours about its probably being cut with baby powder. Finally, when sitting still grew impossible, we ventured outside—first to the George Washington Bridge Bus Station, to watch the heroin addicts nodding off, and the hookers getting lippy with their pimps, and the tired travelers with their entire lives packed up in duffel bags. Then, after the bus station shut down, we headed to Bennett Park, where we slid down the twisting slide and mapped out a terrible short story about two writers in a playground in the middle of the night, high on coke. Ah, the dilettante's-eye view of New York! It's not the recklessness of that memory that surprises me now but the leisure—that we ever had that kind of time to waste, or the freedom and energy to roam. The playground in Bennett Park was fun that night precisely because we weren't meant to be in it. The cops could have booked us for trespassing if they hadn't had real criminals to bust. Now we go to the playground by default, when the kids start bouncing off the walls of our eight-hundred-square-foot apartment.

≡

Sometimes, it's enchanting to play with our kids in the park. I carry bubbles and sidewalk chalk. I fall in love with my husband all over again when he tells an older kid to scram for picking on our son for carrying a doll. We chase our son on his scooter. It can feel like a second childhood. But just as often, it's tedious as hell. We are there as a matter of duty, to kill a weekend afternoon. I injure my wrists from the repetitive stress of pushing the baby *higher, higher*, always *higher*, on the swing. The doctor I consult calls it "mommy wrist" instead of tendonitis. My husband dislikes the helicopter parents at Bennett Park, the gentry of Hudson Heights who intervene when their children grab a toy or push another child, who hover over the rug rats without letting them duke it out on their own, who wear their newborns like accessories. He prefers the playground in J. Hood Wright Park, which resembles the Queens of his childhood, no parents in sight; where the immigrants' kids play pick-up soccer, scream like ambulances, splash in puddles, scarf junk food, forage for mulberries, and peg each other in the head with water balloons. I see what he means, even when our second child gets knocked over by a ten-year-old on a bike. This playground has a wild energy. The kids are free.

J. Hood Wright has a dazzling overlook of the George Washington Bridge, from a platform up a steep flight of stairs where old men in hats sit reading *El Diario* and young men with diamond earrings sit dealing drugs. Every month or so, a film crew arrives to shoot a gritty scene for *Law and Order*, or some other projection of rough New York retrofitted to look like the rotten apple of forty years ago, when it was lousy with crack and crime and the population was slumping.

Nobody calls New York "the Rotten Apple" anymore, except with a kind of nostalgia. More and more, it's referred to as "the playground of the rich," and the population is on the rise. The current baby boom among Manhattan's wealthy class is unique among US cities, with the number of children below age five having grown over 30 percent since the turn of the millennium. My kids are part of this demographic upswing. Their day care borders J. Hood Wright, so we spend a lot of time in its playground. Sometimes a director shouts at us to get out of the movie. *But this is our life*, I want to bark.

The day care is called Bright Beginnings. It costs twelve thousand dollars a year—per kid. That's a discount compared to Bright Horizons, the day care at the nearby medical center, which runs closer to twenty thousand. I now know about the relative costs from the local parents' listserv, which clutters my inbox with alerts of strange men without children spotted in the playgrounds, referrals for lactation consultants, debates about the dangers of vaccination, opinions on when to stop co-sleeping, dissections of the mayor's new policy on universal pre-K, and creative ideas for birthday parties.

When my firstborn's classmate turns two, her parents invite us to her birthday party at Wiggles and Giggles, a playroom on 181st Street and Riverside Drive that rents for $450 an hour. Our son behaves horribly the entire time. "Let's celebrate his birthday at McDonald's," my husband whispers. "Let's not throw him a party at all," I whisper back. We bump fists. While the toddlers eviscerate cupcakes, I look out the window and spot a man in a trench coat holding a revolver. "Is that Liam Neeson?" I ask my husband. Indeed, it is. The actor is playing a hard-boiled detective in a film about seedy New York. When the movie opens, the following year, my husband and I want to see it, but not enough to go through the hassle of finding a babysitter. By this time we have the two children. Day care costs twenty-four thousand dollars. It's been more than a minute since we went on a date. We've lost our old senses of style, the bodies we once wore.

Back when I was single, I traded my dog for a used bike and got a cyclist's-eye view of New York. The bike lanes became a network in my mind, a nervous system. Manhattan was an island whose spine I could navigate in a day, with bridges poking off it like ribs. I got wise to the dangers of car doors opening in my path, the rearview mirrors of city buses at the level of my cranium, and the pleasure of speed on the long thoroughfare of Hudson River Park, with the waterway stretching beside me, my hair whipped back in the wind. I never wore a helmet. My thighs became logs. My rides were epic, and seemingly endless. Then I traded that ride for a stroller.

This is the progressive middle-class parent's eye view of New York

City, beset with anxiety, rumor, distress: We tell ourselves that we didn't choose to live in one of the most diverse cities in the world only to raise our kids in a rarefied bubble. We'd like to send them to a local public school. We believe, as the hallmark of American democracy, that all kids deserve a good education and that private school is not for us. Yet this calculus was far simpler before our own kids arrived. On certain days, we ask ourselves: Why the hell are we raising our children in this quagmire, with its sucky schools and space constraints, its multiple stressors and pollutants, its racing pulse and noise?

Our firstborn is now three, about to turn four, on the brink of entering the public school system. His father and I have toured three pre-school programs so far. Our top choice is a well-regarded community school with eighteen seats in its one pre-K classroom, for which several hundred overeager parents will apply. It has a decent curriculum on Black history, if not on climate change. The crammed open house in the cafeteria smelled of spoiled milk and panic. The parents gathered there were furtively trading tips on "gifted and talented" programs, pipelines to the best high schools. There's a booming test-prep market to get into those kindergartens, which are 70 percent white and Asian, though over 70 percent of the city's public-school students are Hispanic and Black. Ironically, these accelerated programs were meant to promote diversity in urban schools by preventing white flight, but they've effectively resegregated the system. "We're also looking at private schools," one mother confessed. "We're looking at houses in Westchester," a father whispered.

Not since I worked in Gramercy Park has the city seemed more divided. This process is disheartening, to say the least. It turned me into a version of that lady I felt superior to on the park bench, dismissive of the school we are zoned for and anxious to secure a spot in a high-ranked program. Yet when I grow too cynical or stressed about New York's social caste and our children's place within it, it's the playgrounds that save me from wondering how to get back into the garden. I traded my dog and my bike and my wild nights for this: the little ones at play.

Last spring, our boy fell in love with an older girl named Ellia in J. Hood Wright Park. His feeling for Ellia was attached to the poetry of the park, with its tot lot, stegosaurus climber, handball court, dog run, gingko trees, chess tables, outcroppings of schist, picnic area, cigarette butts, sledding hill, ice cream truck, and flagpole, on top of which, on rare occasions at dawn, perches an enormous red-tailed hawk with talons big enough to snatch a squirrel from the rim of a trash can. For my son, who was learning the names of these things, the park contained the spectrum of possibility from menace to joy. The most sparkling of these possibilities was made evident by Ellia's toy tiara.

Ellia's long black hair fell to her waist. Her school uniform was from PS 173, the school we have tacitly agreed, without ever having set foot inside of it, is not good enough for our children. It was easy to see why my three-year-old idolized Ellia. She was pretty and doting, pretending to be his mother, pushing him endlessly on the swing, singing to him in Spanish, and ferrying him on her hip. It was just as easy to see why she cherished my son. He was sweet and adoring, willing to relinquish his snack and follow her commands like a lovesick pup. Their love was mutual and uncomplicated. For a brief spell, they were a couple. One evening at sundown, Ellia led my child out of the playground into the vast park, past the dog run, to the place where the boulders drop off onto the shoulder of the road. I checked an impulse to follow them. She was a third-grader, after all, sensible enough to keep hold of his hand and to watch for broken glass. Soon, they had climbed over the rocks and out of my sight. I didn't know what private magic she meant to show my boy at the park's edge. There is a child's-eye view of New York to which I have no key.

"Geronimo!" I heard Ellia holler. And the peal of his laughter was worth it.

2016

Wilson's Warbler, 1805 Amsterdam Ave., Sugar Hill, Harlem,
muralist: Cara Lynch

OUR NEST

Here's what I can tell you about the Wilson's warbler, which I photographed at a storefront church called Iglesia de Dios El Refugio: They typically avoid the interior of the dense forest, preferring scrubby, overgrown clearings and thin woods; staying low in semi-open areas; and breeding as far north as the timberline, in thickets, second growth, and bogs, along wooded streams, moist tangles, low shrubs, groves of willows, and alders near ponds. They nest from coast to coast in Canada but are far more common in the west. In the Rockies and westward, Wilson's warblers are often the most abundant migrants in late spring. Their young are fed by both parents but brooded by the female alone. Their diet consists mostly of insects, including bees, beetles, caterpillars, aphids, and some berries and spiders. Eight to thirteen days after hatching, young warblers leave the nest. Their nest is usually on the ground, often at the base of shrubs, sunken in sedges or moss, and is shaped like an open cup made of dead leaves and grass, lined with fine grass and hair. Their scientific name is *Cardellina pusilla*. They are known as chipe corona negra, after their black crown, in Mexico, where they overwinter. If global warming keeps apace, 61 percent of their range will be lost.

Here's what I can tell you about our habitat in northern Manhattan: I nest in a rented three-room apartment with my mate and my young on the sixth floor of a mid-rise apartment building with forty units, around the corner from a commuter bus terminal run by the Port Authority and one block from the on-ramp to the George Washington Bridge, in a neighborhood choked by poverty and highways—the Trans-Manhattan Expressway, Henry Hudson Parkway, and Harlem River Drive. We will talk to our young at seven and nine years old, after the murder of George Floyd, about how to protect themselves from the police, who disproportionately apprehend Black and brown youth in our habitat. Among our kind, this grim warning is known as "the Talk." The walls of our nest are covered in artwork collected by us or made by our young. Our young share a bunk bed in a room that fits little else. Our nest is blessed by stacks of books, the sounds of merengue and bachata outside on the street, western light, views of the rooftops and parapets of the buildings around us (studded

View from 804 W. 180th St., Apt. 62,
Washington Heights, Manhattan

with boiler pipes belching black smoke from heating oil No. 4), and the wide-open sky over the Hudson River, which we cannot see.

In the long hallway cluttered with shoes and boots is an umbrella stand with three janky umbrellas and a rolled-up poster of Frederick Douglass which sometimes our young unroll and say looks like Grandpa and other times leave rolled up like a tube to use as a telescope or a sword. Our diet includes Chinese takeout from Great Wall, roast chicken from El Malecon, and pupusas from Mi Paso Centroamericano. On Halloween, our young eat candy they forage by trick-or-treating from the mom-and-pop shops along the business thoroughfare of Broadway. Across the street from the bus terminal is the church where we married, where we baptized our young, and where an undocumented neighbor named Amanda Morales took refuge for months with her American-born children as part of the sanctuary movement, to avoid deportation to Guatemala by federal immigration agents.

Two blocks north of the Wilson's warbler mural on Amsterdam Avenue sits an office of WE ACT for Environmental Justice, an organization that empowers low-income people of color to advocate for healthy communities. It is well documented that some of the most polluted environments in the nation are where people of color live, work, play, and pray. For instance, a recent environmental health report on northern Manhattan by WE ACT stated that, as a result of risks including poor air quality from heavy diesel-truck traffic, dirty boilers, power generation plants, and polluting municipal infrastructure like bus depots and sewage treatment plants, Harlem has a childhood asthma hospitalization rate six times the national and three times the citywide average. Another recent report from the NAACP and the Clean Air Task Force showed that Black Americans are 75 percent more likely than other Americans to live in fence-line communities sited near facilities producing hazardous waste. A recent study by the Environmental Protection Agency (EPA)'s National Center for Environmental Assessment found that regardless of their income level, Black Americans are exposed to higher levels of air pollution than white Americans—1.5 times as much of the pollution from burning fossil fuels as the population at large.

I am the mother of Black children in America. It's not possible for me to consider the threats posed to birds without also considering the threats posed to us.

2020

Protect Yo Heart, Upper Manhattan,
artist: Uncutt Art

KNOW YOUR RIGHTS!

On the Saturday after the Charleston church massacre, when nine worshippers at one of the nation's oldest Black churches were slaughtered during Bible study by a white gunman hoping to ignite a race war, we dragged our kids to the east side to walk them over New York City's oldest standing bridge. It seemed as good a way as any to spend a weekend afternoon, and a refreshing change from J. Hood Wright Park.

The High Bridge, which was built with much fanfare in the mid-nineteenth century as part of the Croton Aqueduct system and as a promenade connecting Upper Manhattan to the Bronx over the Harlem River, had recently—and somewhat miraculously—reopened after forty-odd years of disuse. I say "miraculously" because the bridge was a piece of infrastructure most of us had come to accept as blighted, even as some civic groups had coalesced to resurrect it. In the back of our minds that summer of 2015, as an uprising and its violent suppression reverberated in Missouri, was the problem of when and how to talk to our children about protecting themselves from the police.

At what age is such a conversation appropriate? By what age is it critical? How could it not be despairing? And what, precisely, should be said? Our eldest was four then; our youngest, just two.

The day was hot. En route to the bridge, we felt no reprieve from the sun, just as we'd felt no relief from the pileup of bad news about Blacks being murdered with impunity. When we learned of the terror at Mother Emanuel AME in Charleston, we had not yet recovered from the unlawful death of Freddie Gray in Baltimore, or the shooting of Mike Brown in Ferguson, or the strangling of Eric Garner in Staten Island, or the shooting of Trayvon Martin in Florida, or the shooting of Tamir Rice in Cleveland, to name but a few triggers of civil unrest. We weren't surprised there were no indictments in these cases, sadly enough, but we were righteously indignant. The deaths seemed to be cascading in rapid succession, each one tripping a live wire, like the feet of Muybridge's galloping horse.

The picture we were getting—and not because it was growing worse, but because our technology now exposed it—was clear and mounting evidence of discriminatory systems that don't treat or protect our citizens equally, and escalating dissent was giving rise to a movement that insists what should be evident to everyone: Black lives matter. There were hashtag alerts for pop-up protests in malls, die-ins on roads, and other staged acts of civil disobedience, such as disruptions of white people eating their brunch. Protesters against police brutality dusted off some slogans from the civil rights era, such as "No justice—no peace!" but others were au courant: "I can't breathe," "Hands up, don't shoot!" "White silence is violence," and, most poignant to me as a mother, "Is my son next?"

"It's too *hot*, and my legs are too small," our firstborn protested on the way to the bridge.

The boy was right—it was hot, and getting hotter. He was tall for four but still so little. When he stood at our front door, his nose just cleared the height of the doorknob. He was the same size as the pair of children depicted in a two-panel cartoon by Ben Sargent that circulated widely on my Facebook feed that summer. Both panels depict a little boy at the threshold of his home, on the verge of stepping outdoors. The drawings are nearly identical except that the first boy is white and the second, Black. "I'm goin' out, Mom!" each boy calls to a mother outside of the frame. The white boy's mother simply replies, "Put on your jacket." But the other mother's instructions comprise so intricate, leery, and vexed a warning that her words obstruct the exit: "Put on your jacket, keep your hands in sight at all times, don't make any sudden moves, keep your mouth shut around police, don't run, don't wear a hoodie, don't give them an excuse to hurt you . . ." and so on, until the text in her speech bubble blurs into an illegible cipher, as in a painting by Glenn Ligon. The cartoon is titled *Still Two Americas*.

I didn't wish to be her, the mother who needed to say, "Some people will read you as Black and therefore X." Why should I be the fearful mother factoring for X? Nor did I covet the white mother's casual regard. I wanted to be the mother who got to say to her children, "Stay curious," as well as "Enjoy yourselves!" on their way out the door.

But for now, I carried our sweaty second-born down 173rd Street on my back, while my husband led our stubborn firstborn by the hand. You know the thermometer's popping in Washington Heights when there aren't any Dominicans out on the sidewalks playing dominoes. Nobody had yet cranked open the fire hydrants. The heat knocked out the two-year-old as if it were a club. The four-year-old was in a rotten mood. He demanded a drink, then rejected the water bottle we'd packed. He whined that the walk was too long, then challenged our authority in a dozen other hectoring ways until we at last arrived at Highbridge Park. There he refused to descend the hundred stairs to the bridge by flinging himself onto the asphalt, with his arms and legs bent in the style of a swastika, not five feet from a dead rat. The kid's defiance bothered us for all the usual reasons a parent should find it irksome but also because, if allowed to incubate in the ghetto where we live, that defiance could get him killed. What if he got lippy with a cop? When would they stop seeing him as a child and start seeing him as a problem? Seventeen, like Trayvon Martin? Fourteen, like Emmett Till? Twelve, like Tamir Rice? If I'm honest, that narrative had already begun, in preschool even before we started stressing about kindergarten placement.

They told us that he was "noncompliant." That he was "difficult." That he didn't want to stay in his seat. He didn't want to learn about math unless it was clear to him how the math would be used. They said he was "aggressive." That he was "a problem." That he didn't want to follow the rules unless the rules made sense. That he wanted to read his book about trains rather than listen at story time. That he had a deficit of attention, a disorder that could be curbed with meds. They said nothing about his intelligence.

Our son was soon coaxed down the vertiginous stairs by the magical horn of an Amtrak train on the railway beneath the bridge. He has explained to me his fierce attraction to trains and boats and vehicles in general with irritation that I didn't already know the reason: "They take you somewhere else." That's just it. From the time your children begin walking, they are moving away from you. This is as it should be, even when you can't protect them from harm with anything but the inadequate quilt of your love.

A sweet old man in seersucker shorts stopped us at the entrance to the bridge to make sure we appreciated the marvel of its rehabilitation. He was something of a history buff and spoke with a Balkan accent—Greek, I think. He could recall when the bridge was shut down, after its long fall into decline, and the time before that when miscreants and vandals tossed projectiles over the guardrail into the polluted water below or at the traffic on Harlem River Drive. Thanks to him, I know that the bridge was a feat of engineering originally modeled after a Roman aqueduct, siphoning water from Westchester County through pipes beneath its walkway into the city, enabling New Yorkers to enjoy their first indoor plumbing (including the flush toilet). The old man never thought he'd live to see the day when the High Bridge was back in business, and he was proud that the citizen-led campaign to reopen it had succeeded. "This bridge changed everything," the old man said in wonderment, as if the relic was a truer paean to empire than the skyscrapers twinkling in the skyline far to the south of us—the Chrysler Building, the former Citicorp tower, and the spire of the Empire State.

Dutifully, we paraded across the Harlem River to the Bronx. Maybe it was because I so admired the old man's perspective, attuned as it was to a less conspicuous wonder of the world, that on our return trip, I noticed a mural I could have sworn had not been there before.

"KNOW YOUR RIGHTS!" the mural trumpeted in capital letters. How had it escaped my attention? The artwork covered a brick wall abutting the twenty-four-hour laundromat I passed every weekday morning on the walk to the children's day care. A vision in tropical blues, the same tones as the baby quilt sewn by my mother for our firstborn, it splashed out from the gritty, gray surroundings, creating an illusion of depth. I loved it as I would come to love the birds. My eyes drank it in.

This mural operates like a comic strip, in panels marrying image and text. In the first panel, a youngster is carded by a law enforcement official. In the second, a goateed man in a baseball cap is being handcuffed. In the third, a group of citizens stare evenly outward. One of them wears a look of disgust and a T-shirt that says, "4th Amendment," a sly allusion to

Know Your Rights mural, Wadsworth Ave. and 174th St., Washington Heights, Manhattan, Nelson Rivas, aka Cekis, 2009. "If you are detained or arrested by a police officer, demand to speak with an attorney and don't tell them anything until an attorney is present."

the part of our constitution that protects us against unreasonable search and seizure without probable cause. Another holds his cell phone aloft to record what is happening on the street. "You have the right to observe and film police activity," the mural states in Spanish, appropriate for a neighborhood where Spanish is the dominant language and where young men of color are regularly stopped and frisked by the police. In the lower left-hand corner, the Miranda rights are paraphrased in English.

My first instinct was to take a picture of the mural so that I could carry it with me in my pocket. I was grateful for it, not only as a thing of beauty on the gallery of the street but also as a kind of answer to the question that had been troubling us—how to inform our children about the harassment they might face. The mural struck me as an act of love for the people who would pass by it. I understood why it had been made, and why it had been made here in the hood next to a laundromat, as opposed to on Fifth Avenue next to Henri Bendel, Tiffany & Co., or Saks. It was armor against the cruelty of the world. It was also a salve, to reclaim physical and psychic space. I wondered who had made it.

After some Internet sleuthing, I discovered the painter was a Chilean artist who goes by the tag name Cekis, and that this mural was the first of several public artworks commissioned by a coalition of grassroots organizations called Peoples' Justice for Community Control and Police Accountability. The other Know Your Rights murals were spread out across four of New York City's five boroughs (excluding Staten Island, where a great number of cops live), in poor neighborhoods most plagued by police misconduct, like ours. For the rest of that summer and into the fall, I photographed as many of them as I could, like a magpie collecting bright things for her nest. But also, searching for the murals was freedom from the nest itself, a return to the traveler I knew myself to be before having the kids clipped my wings. My old way of knowing the world was lost, but this was a way to regain it—or rather, to reframe it with a new way of seeing.

As with the mural in Washington Heights, when I shot this mural in Harlem, I chose to capture a passerby in the frame to give a sense of scale, but with the intent to preserve the subject's anonymity. Thrown against a sharp white background, the man in Harlem appears in silhouette, his beard like Thelonious Monk's, his shadow extending from his feet, and the shadow of the fire escape above him slanting down against the mural like the bars of a cage. A woman depicted in the mural's foreground holds a bullhorn to her mouth. A portion of the text reads, "Write down the officer's badge name, #, and/or other identifying info. . . . You don't have to answer any questions from police." Her advice is specifically targeted to those at risk of being stopped and frisked.

Stop-and-frisk policing was implemented in New York as part of a trend of increased enforcement that began in response to rising crime and the crack cocaine epidemic of the 1980s and '90s. The technique disproportionately affects young men of color. From 2004 through 2012, the years between starting my job and becoming a mom, African Americans and Hispanics were subject to nearly 90 percent of the 4.4 million stop-and-frisk actions taken by police officers, despite constituting only about half of the city's population. In Black and Latino neighborhoods

Know Your Rghts mural, 138th St. and Adam Clayton Powell Jr. Blvd., Harlem, lead
artist: Sophia Dawson, 2013. "Write down the officer's badge #, name, and/or other
identifying info." "Get medical attention if needed and take pictures of injuries."
"You don't have to answer any questions from police. When they approach, say, 'Am I
being detained, or am I free to go?' If they detain you, stay silent + demand a lawyer.
A frisk is only a pat down. If police try to do more than that say loudly, 'I do not
consent to this search.'" "You have the right to observe, photograph, record, and film
police activity."

like Harlem and Washington Heights, residents often view the police, a
force ostensibly there to protect them, with mistrust and fear. In 2013,
the year the Harlem mural was made and my second son was born, a
federal court judged the use of stop-and-frisk tactics by the NYPD to be
excessive and unconstitutional. Since then, their use has declined. Critics
of reducing the practice predicted a rise in crime; instead, overall crime
has dropped. I would like to believe these statistics mean it's growing
slightly safer for my children to walk.

Yul-san Liem, who works for one of the activist organizations that
make up Peoples' Justice, explained to me in a phone call during my

kids' naptime that the murals were financed by the Center for Constitutional Rights. "Visual art communicates differently than the written or spoken word," she commented. "By creating Know Your Rights murals, we seek to bring important information directly to the streets where it is needed the most, and in a way that is memorable and visually striking."

Peoples' Justice formed in 2007 in the wake of the NYPD killing of the unarmed Black man Sean Bell the day before his wedding. "It wasn't an isolated incident," Liem lamented, recalling the 1999 killing of Amadou Diallo, the unarmed Black man shot forty-one times by police, and the assault of Abner Louima, who in 1997 was sodomized by police with a broomstick and allegedly told, "Take this, nigger!" Liem said, "Our original goal was to highlight the systemic nature of police violence in communities of color. We've taken a proactive approach to empowerment that includes organizing neighborhood-based Cop Watch teams and outreach that uses public art as a means of education. It's about shifting culture and creating hope."

Maybe that's what I was scavenging for: hope. I like how Emily Dickinson defined it—"the thing with feathers."

I had difficulty finding the third mural that I shot in Bushwick, Brooklyn, in part because Bushwick is a neighborhood of multiple murals but also because Liem had given me bum directions. I lost myself in the rainbow spectacle of street art. There was Nelson Mandela on a wall overlooking the parking lot of a White Castle, but where was the mural I sought? I asked a group of kids in Catholic-school uniforms if they knew where I could find it. They all claimed to know the Know Your Rights mural, but none could give me an exact address. Either it was somewhere down Knickerbocker Avenue or else it was in the opposite direction past three or four schoolyards and a car wash. In the end, one girl kindly volunteered to walk me there. She wore a purple backpack, braces on her teeth, and a gold name necklace that said, coincidentally (or not), "Esperanza." Esperanza told me with excitement that she'd be getting an iPhone like mine for her thirteenth birthday. After perambulating for a half an hour, we finally located the mural in an overgrown lot behind a chain-link fence. We'd had so many false sightings at that point that I sensed it was part of the girl's mental, rather than physical, landscape.

Know Your Rights mural, Irving and Gates Aves., Bushwick, Brooklyn, artist: Dasic Fernández, 2011. "If you are harassed by police, write down the officer's badge number, name, and/or other identifying information. Get medical attention if you need it and take pictures of any injuries." "All students have the right to attend school in a safe, secure, non-threatening and respectful learning environment in which they are free from harassment." "No tenant can be evicted from their apartment without being taken to housing court." "Owners are required by law to keep their buildings safe, well maintained and in good repair. If not, call 911."

It rose out of the weeds in pastel shades like an enormous Easter egg. "I love this one," she confessed. "It's so *big*."

As with the murals in Washington Heights and Harlem, the text of the Bushwick mural exhorts the viewer to watch and film police activities. This time the message is underscored by a figure in the foreground who points to her enormous eye as if to say, "Watch out. Keep your eyes open." I wanted an eye that big. A man directly behind her uses his phone to film a police officer making an arrest in the mural's background. The phone is configured as a weapon for social change. The teenager that I photograph walking past the mural is also on the phone. Though she appears oblivious to the mural she also appears, in the context of my photo, to be wielding a tool. That is, the phone distracts her from being present, but she could also deploy its camera at any moment to record what's happening on the street.

The fourth mural that I shot was painted on a corrugated fence in Long Island City, Queens, across from the Ravenswood housing projects. On my way through the projects, I passed a barefoot lady in a fabulous church hat pushing a stroller full of cans. She was involved in a heated argument over a MetroCard with a man invisible to me. In the middle of an invective, she stopped to tell me he was a lying thief. "I believe you," I said, emphatically. "He's a jerk." We smiled at each other. She returned to her dispute, and I went on my way.

"If you are harassed by police," the Long Island City mural advises, "take pictures of any injuries." Again, the mural forms a backdrop for the people walking by—but in this case, because it consists entirely of text,

Know Your Rights mural, 35th Ave. and 12th St., Long Island City, Queens, artist: Dasic Fernández, 2012. "If you are HARRASSED [sic] by police, write down the officer BADGE number, name and/or other identifying information. Take PICTURES of any INJURIES."

the message is even starker. A woman is about to cross the street. I don't know where she's going or what she's looking at; she may be checking for oncoming traffic or reading the warning on the mural. Her braids swing across her back as her sneaker approaches the curb. My friend Garnette Cadogan has written, "Walking is among the most dignified of human activities." But here, the woman's simple, dignified act of walking, whether home from work or school or to the bodega for a carton of milk, is erupted by the somber memo that hangs in the background. The public space feels contested and even traumatic because of the public art. The intersection looks hazardous, like something is about to hit her.

That's how it felt when my father gave me and my brothers the Talk—like something was about to hit me, and my childhood was over. I was ten. We had just finished dinner. My mother was clearing the plates to wash the dishes while tending my baby brother, who was one. My older brother was thirteen; my younger brother was seven. Our father sat at

W. 183rd St. and Broadway, Washington Heights, Manhattan, muralist: Timothy Goodman, 2018. "I WANNA BE AS HUMAN AS POSSIBLE TO NOT HIDE RUN DESTROY OR PROVE ANYTHING BUT TO SEE AND BE SEEN."

the head of the table. He had poured himself a glass of Johnnie Walker Black Label, on the rocks, with a splash of sweet vermouth and a twist, to steel himself for the necessary lecture he had to deliver.

He told us that US democracy had never been a true democracy; that as far as nations go, America was still a child; that growing up here would be tough. Because of white supremacy, some people would think negatively of us, no matter how smart we were, no matter how poised, how well-dressed, well-spoken, or well-behaved. We would have to work twice as hard to get half as far. There were different rules for Black people and, in particular, for young Black men. The streets were not as safe for my brothers as they were for their white friends. Society was framed to see them as menacing and even disposable. They would be profiled and mistrusted. My brothers, whose skin was darker than mine, already had two strikes against them and a target on their backs, but that didn't mean things would be easier for me. As a female, I would be preyed upon by men who would try to take advantage of me, and as a light-skinned Black girl who could pass for white, I would hear white people spewing all kinds of ugly untruths about Black folks, whose dignity I would have to defend. He didn't want to have to give us this talk, which his own mother hadn't wanted to give him. She had delayed giving him the Talk because she didn't want him growing up hating white people, but at a certain age he had needed to know that the consequence of questioning authority could be death, and now, so did we.

Then, without going into all the details, he told us a white man had killed our grandfather and had gone unpunished for the crime. Despite this history, we deserved to question authority, and furthermore, we deserved to exist freely in our skin. The world needed our bodies in it. We were full of promise. We had a bright future. Our Blackness wasn't a

[OPPOSITE] Know Your Rights mural, Marcus Garvey Blvd. and MacDonough St., Bedford-Stuyvesant, Brooklyn, artists: Trust Your Struggle (collective), 2010. "Justice or Just Us?" "LOVE/HATE." "Stay calm and in control. Don't get into an argument. Remember officer's badge and patrol car number. Don't resist, even if you believe you're innocent. You don't have to consent to be searched. Try to find a witness & get their name & contact. Anything you say can be used against you. Know Your Rights. Trust Your Struggle. Spread love. It's the Brooklyn way. Didn't pass the bar, but know a little bit; enough that you won't illegally search N.Y."

life sentence but a gift. Our people had survived unspeakable atrocities and shaped history. Our ancestors kept faith, made music and community, found love and joy. Our culture was joyful. We belonged to a loving Black family. We should be cautious, but more than that, we should be grateful for our heritage.

After the Talk, our father poured himself another drink to loosen up and, without explanation or apology, recited the first eighteen lines of the *Canterbury Tales* by heart. Maybe he meant to clear the air.

> . . . (So priketh hem Nature in hir corages),
> Thanne longen folk to goon on pilgrimages,
> And palmeres for to seken straunge strondes,
> To ferne halwes, kowthe in sondry londes; . . .

What did it all mean? I remember sitting at the table with goose bumps because of the way his words circumscribed themselves on my skin. Sometimes I think my entire life since that night has been an effort to connect the dots between those two messages—the Talk and Chaucer's prelude. I longed to go on pilgrimages in defense of our very life.

Bedford-Stuyvesant is a swiftly gentrifying Brooklyn neighborhood made famous by Spike Lee's landmark film *Do the Right Thing*. In fact, the Bed-Stuy mural directly references that movie by depicting the character Radio Raheem. At the start of the movie, Radio Raheem blasts Public Enemy's "Fight the Power" from his boom box like a reveille. Near the movie's end, he's choked to death by a nightstick-wielding cop—a pivotal plot point that incites a riot, much like the uprisings that followed the Rodney King verdict in Los Angeles, and the Freddie Gray verdict in Baltimore, and the Michael Brown verdict in Ferguson, which reverberated across the country like so many waves of heat.

In New York, after Ferguson, I remember the protesters took to the streets chanting, "Whose streets? Our streets!" We were advised by activists in Palestine resisting state violence half a world away to always run against the wind, to keep calm when tear-gassed, and not to rub our eyes; the pain would pass. I myself was drawn to the vortex of 125th Street, where I shot pictures of the crowd swarming toward the Triborough Bridge. I paused there at the edge of my own reason, sometime before midnight, to return to my children in the glow of the nightlight, tucked beneath the baby quilts sewn by my mother in their little beds in the bedroom we'd painted green like the green of *Goodnight Moon*, with decals of tree trunks on the walls so it looked like the room was blurring into a forest, the dollhouse under the fire escape window, the stuffed elephant from Uganda on the radiator, the four-year-old in his toddler bed and the two-year-old on a sleep mat under their father's former writing desk, now rigged with curtains that opened and closed like a puppet theater, repurposed from the shalwar kameez I'd bought in Brooklyn's Little Pakistan and worn to my baby shower where one of the women—it could have been any one of them—said that motherhood would feel like my heart was walking on the outside of my body. I probably sang the lullabies I sang in those days—"Tura Lura Lura," the Irish lullaby my mother had sung to me; "Edelweiss"; and the song I made up about a train full of all the people my children knew and loved: their cousins, grandparents, aunts and uncles, friends, teachers, neighbors, the librarian, the super, the mailman, the lady in apartment 1E and her two little dogs, and on and on, packing the cars of the train with every blessed name they could supply until they fell back asleep.

Meanwhile, the rebels pushed on as far as the tollbooths on the Manhattan side, succeeding in shutting the bridge down. It felt so logical an impulse, to act unruly in the face of misrule. Yet this impulse is what the Bed-Stuy mural admonishes against.

Radio Raheem's fist is the focal point of the mural, adorned with its gold "LOVE" knuckle plate. The mural, dominated by the color red, cautions the viewer to "Stay calm and in control. Don't get into an argument. . . . Don't resist, even if you believe you're innocent."

The man I photograph walking past the love punch wears paint-splattered work boots, a headcloth over his dreadlocks, and earphones. I wonder what he's listening to. Perhaps because he's distracted by his music, he's unaware that I've shot him with my phone.

≡

So was the woman in the Bronx, where I took my sixth and final picture. She was too absorbed by the screen of her device to notice me, though if she had looked my way, she would have seen that I, too, was operating my phone. My posture mirrored the person in the mural, who films a plainclothes police officer cuffing a man over the hood of a car. I had to wait over an hour to get this shot because a belligerent man who didn't wish to be photographed nevertheless refused to get out of the frame. After pissing on the sidewalk, he zipped up and eventually drifted off beneath the Bruckner Expressway. Of all the neighborhoods I traversed, Hunts Point felt the roughest, by which I mean the most neglected. On the long walk to the Point from the elevated 2 train, through the red-light district, I was surveyed with interest. I felt that if I wasn't mindful, someone down on their luck might succeed in snatching my phone. Yet I stayed planted by the mural, looking for something concrete. Why? Because I was afraid.

The phone in the Hunts Point mural is almost as tall as the woman walking beneath it, its screen the approximate size of her handbag. On the screen we see repeated the nested image of the plainclothes police officer cuffing a man over the hood of the car. The dizzying effect of the mural is to put the viewer in the perspective of the photographer.

I have fallen into the mural—or rather, the mural has sucked me in. I am the third dimension, the watcher. I am the photographer with the

Know Your Rights mural, Barretto St. and Garrison Ave., Hunts Point,
Bronx, artist: Dasic Fernández, 2012. "You have the right to watch &
film police activities." "If you are detained or arrested by a police officer,
demand to speak with an attorney and don't say anything until attorney is
present."

phone in her hand. So, potentially, is the passerby, though in this context
her posture is also a reminder that passivity has its cost. The woman is
about to step out of my frame. For now she is caught, as in a web, by the
shadows of power lines and trees. The text behind her echoes that of
the first mural I shot, on the streets of Washington Heights: "If you are
detained or arrested by a police officer, demand to speak with an attorney
and don't say anything until attorney is present."

It was as if the text was on a loop. I'd begun to feel I was moving in
circles. I had a personal stake in this documentation, as the wife of a
Black man and mother of two small Black boys. Our kids were still too
young for the Talk, but it was on our horizon. In a sense, I'd already been
preparing for it, ever since the night the Talk was delivered to me. The
murals offered loving instructions to add to what I'd already been told.

I stopped to take stock of my pictures, scrolling backward. Though the style of each mural was distinct, the message was the same: somebody loves you enough to try to keep you safe by informing you of your rights. The murals' insistence on those rights, which the citizens of our nation don't yet equally enjoy, reminded me that, like the High Bridge, the Constitution was just another lofty piece of infrastructure in need of rehabilitation. Such changes do occur, it seems. Were it not for the fact that I shot them in different locales, I felt I could craft a zoetrope of the passersby to show my children. The many walkers would appear unified as one—even if at times that walker was a woman or a man, Black or brown, old or young—advancing toward one steady goal. "Look how marvelous," I would say of the moving image. And if my children asked me where the walker was going, I would answer, "To the bridge."

2015–2016

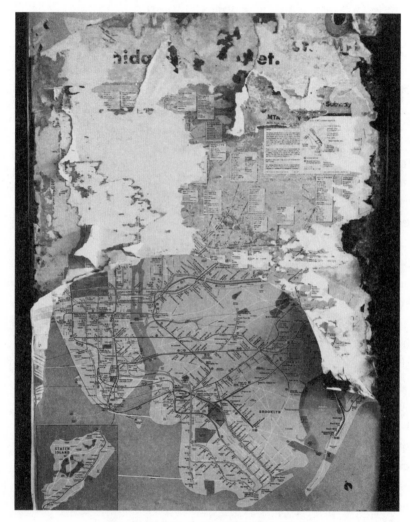

How Do We Get to the Guggenheim from Here?,
23rd St., Chelsea, Manhattan, downtown C train platform

CLIMATE SIGNS

For Mik

Our son's love of trains was once so absolute I never foresaw it could be replaced. New York City is a marvelous place to live for train-obsessed boys. When he was three and four, we spent many a rainy day with no particular destination, riding the rails for the aimless pleasure of it, studying the branching, multicolored lines of the subway map, which he'd memorized like a second alphabet. I'd hoist him up to watch the dimly lit tunnel unfurl through the grimy front window of the A train's first car as it plunged us jerkily along the seemingly endless and intersecting tracks. Some rainy mornings, our destination was Eighty-First Street, where we exited the B or C with dripping umbrellas and his little sibling in tow to enter the American Museum of Natural History.

There, at a special exhibition called Nature's Fury, our son's attention turned like a whiplash from trains to violent weather. Even before this show, the museum demanded a certain reckoning with the violence of the Anthropocene. What grown-up wouldn't feel a sense of profound regret confronting the diorama of the northern white rhinoceros in the hall of African mammals, or the hall of ocean life's psychedelic display of the Andros coral reef as it looked in the Bahamas a century ago? Meandering the marble halls of the natural history museum is like reading an essay on losing the earth through human folly. Yet none of its taxonomies of threatened biodiversity, not even the big blue whale, moved my kindergartner like Nature's Fury.

The focus of the immersive exhibition was on the science of the worst natural disasters of the last fifty years—their awesome destructive power and their increasing frequency and force. Accompanied by a dramatic score of diminished chords and fast chromatic descents, the exhibit meant to show how people adapt and cope in the aftermath of these events, and how scientists are helping to plan responses and reduce hazards in preparation for disasters to come.

"Are they too young for this?" my husband questioned, too late. Our impulsive boy had darted ahead and cut the line to erupt a virtual volcano. I supposed it made him feel less doomed than like a small god that, in addition to making lava spout at the push of a button, the kid could manipulate the fault lines of a model earthquake, set off a tsunami, and stand in the eye of a raging tornado.

In the section on hurricanes at a table map of New York, our eldest was also able to survey the sucker punch that Hurricane Sandy delivered to the five boroughs. This interactive cartography was a darker version of the subway map he'd memorized, detailing the floodplains along our city's 520 miles of coast. I can still see my boy there, his chin just clearing the table's touch screen so that his face was eerily underlit by the glow of information, while our youngest crawled beneath. Seventeen percent of the city's landmass flooded, leaving two million people without power, seventeen thousand homes damaged, and forty-three people dead. On the map, the water rose to overtake the shorelines at Red Hook, Battery Park, Coney Island . . . All across the Big Apple, the lights were going out.

"Come away from there," one or the other of us called uneasily, because we weren't prepared to confront what climate change would mean for our children, to say nothing of our children's children. The firstborn was five at the time; the second-born was three. In their lifetimes, according to a conservative estimate in a recent report by the New York City Panel on Climate Change, they could see the water surrounding Manhattan rise six feet. We pulled them away from that terrifying map of our habitat to go look at dinosaur bones—an easier mass extinction to consider because it lay in the distant past.

What strikes me now as irrational about our response then isn't our ordinary parental instinct to protect our kids from scary stuff—it's our denial. Their father and I treated that display as a vision we could put off until later, when it clearly conveyed what had already transpired. Structural racism predated our climate concern and, even now, governs it, like a preexisting condition.

"We are now faced with the fact, my friends, that tomorrow is today. We are confronted with the fierce urgency of now," preached Martin

Luther King Jr. in 1967, in one of his lesser-known sermons, "Beyond Vietnam: A Time to Break Silence." He may as well have been speaking on climate change. Sandy made landfall in 2012, the year after the boy was born, while I was pregnant with his brother. It gave a preview of what the city faces in the next century and beyond, as sea levels continue to rise due to melting ice sheets. The storm exposed our weaknesses, and not just in the face of flooding: I remember that when the bodegas in our hood ran out of food, some folks shared with their neighbors. But when the gas station started running out of fuel, near the dual-language public school our kids now attend, some folks pulled out their guns.

As much as we may worry about our kids' future, it's already here.

Avoiding the map didn't annul its impact on our son. The subject of storms had gripped his consciousness as surely as his author father's had been gripped by horror films. That part of the boy's brain that previously needed to know the relative speed of a Big Boy steam engine to a Shinkansen bullet train now needed to know what wind speed differentiated a Category 4 hurricane from a Category 5. Soon enough, and for months afterward, Mr. Wayne, the friendly librarian at the Fort Washington branch of the New York Public Library, would greet our boy with an apology: "I'm sorry, son." (Thank God for the folks in our community who claim him as kin.) There were no more books in the children's section on the subject of violent weather besides those he'd already consumed.

At bedtime, while my second-born sucked his thumb to sleep, I offered my firstborn reassurance that we weren't in a flood zone; that up in Washington Heights—as the name suggests—we live on higher ground. "You're safe," I told him.

"But the A was flooded during Sandy," he reminded me, matter-of-factly. "The trains stopped running, and the mayor canceled Halloween." Then he went on rapturously about the disastrous confluence of the high tide and the full moon that created the surge, while I tried to sing him a lullaby.

Eventually, a different fixation overtook extreme weather, and another after that. Such is the pattern of categorical learners. It may have been sharks before the *Titanic*, or the other way around—I've forgotten. Two years have passed since we saw Nature's Fury; a year and a half since our president led the United States to withdraw from the Paris climate accords. The boy is seven now, what Jesuits call "the age of reason." His brother is five and learning to read. If current trends continue, the world is projected to be 1.5 degrees Celsius warmer than preindustrial levels by the time they reach their late twenties. The scientific community has long held two degrees Celsius to be an irreversible tipping point. Two degrees of global warming, according to the UN's Intergovernmental Panel on Climate Change (IPCC), marks climate catastrophe.

At two degrees, which is our best-case climate scenario if we make seismic global efforts to end carbon emissions, which we are not on course to do, melting ice sheets will still pass a point of no return, flooding NYC and dozens of other major world cities; annual heat waves and wildfires will scrub the planet; drought, flood, and fluctuations in temperature will shrink our food supply; water scarcity will hurt four hundred million more people than it already does. Statistical analysis indicates we have only a 5 percent chance of limiting warming to less than two degrees. Two degrees has been described as "genocide."

In fact, we're on track for over *four* degrees of warming and an unfathomable scale of suffering by century's end. By that time, if they're lucky (and have had no fatal encounters or illness), our children will be old. It's pointless to question whether or not it was ethical to have them in the first place since, in any case, they are here. Their father writes about imaginary horrors. For my part, I'm only beginning to see that the question of how to prepare our kids for the real horrors to come is collateral to the problem of how to deal as adults with the damage we've stewarded them into. We already knew we needed to prepare them for Black life under white supremacy. Now I see we also have to prepare them for extreme heat and ungodly flooding.

What helped me to see this was a road sign. I came across it in fall 2018 in Harlem's St. Nicholas Park, two weeks before the release of the United Nations climate report that concluded we must reduce greenhouse gases to limit global warming to the 1.5-degree threshold. The sign was part of another exhibit, but I didn't know that when it stopped me in my tracks on my way to work. It was one of those LED billboards you normally spot on a highway, alerting drivers to icy conditions, lane closures, or other safety threats ahead. Oddly enough, the sign was parked in the grass two-thirds of the way up the vertiginously steep slope to City

College. *How did that get there?* I wondered. More surprising than the traffic sign's misplacement was its message:

CLIMATE DENIAL KILLS

St. Nicholas Park recently ranked among NYC's top five most violent parks, as measured by its high rate of crime. I was assaulted there once by a girl in a gang who coldcocked me in the face. I'd been walking there with my husband, back when we were still dating, caught up in a conversation about his defaulted student loan debt, when—*bam!*—all of a sudden I felt like I'd roller-skated into a tree. For a second, I blacked out. Neither of us had seen the punch coming, though he had thought it strange that one of the girls in the group crossing our path had pointed at me with a cane. Maybe the other girl who then hit me was an initiate, and I was her mark? Was it because she mistook me for white? We were genuinely bewildered about how to respond. We weren't about to put those girls at risk by calling the cops; one of them looked pregnant. Now they were getting away. He asked if I wanted him to go after them and retaliate. *No,* I said, *just find me some ice.*

This sign hit me almost as hard. I felt as if someone had punched through from another dimension to shock me awake. And I felt just as confused about how to respond. Was I seeing the sign correctly? Yes. It repeated its declaration in Spanish:

LA NEGACIÓN CLIMÁTICA MATA

Every couple seconds, the sign refreshed, unspooling a disquieting, if strangely droll, string of warnings:

NO ICEBERGS AHEAD
50,000,000 CLIMATE REFUGEES
CAUTION
CLIMATE CHANGE AT WORK
ABOLISH COAL-ONIALISM

and so on.

The familiar equipment of the highway sign gave authority to the text. Because it was parked in the wrong place, the sign appeared hijacked—as

in a prank. I understood myself to be the willing target of a public artwork but not who was behind it. The voice was creepily disembodied. I
admired its combination of didacticism and whimsy, but even with its
puns, the sign was more chilling than funny. The butt of the prank was
our complacence, our lousy failure to think one generation ahead, let
alone seven, as is the edict of the Iroquois' Great Law.

For several minutes, I paid humble attention to the sign, unsure how
to react. The only practical guidance it offered was to "VOTE ECO-
LOGICALLY"—something achievable, given the upcoming midterm
elections. But what else was the sign telling us to do? My individual prac-
tices of composting and giving up plastic bags felt lame when the headlines
were warning of genocide and civilization's end. I began to feel exposed,
standing there, and briefly considered that I might be on *Candid Camera*.

I looked around the park for help, half hoping the artist might pop
out from behind a tree to explain him- or herself. I wanted to process
the work's messaging with somebody else. It signaled the tip of a melting
iceberg whose magnitude surpassed my cognition. How to move past the
paralyzing fear that whatever we do is too little, too late? The trouble here
was one of scale. Sadly enough, no one else in the park that morning—
not the dog walkers, or the City College students, or the man sleeping in
the grass—seemed engaged by the sign at all. And so, I snapped a picture
of it with my iPhone and shared it on Twitter.

Almost immediately, a stranger with the handle @AwakeMik replied
with a photo of an identical sign he'd just discovered in Sunset Park,
Brooklyn, while walking his dog, Chester, named after Himes. "VOTA
ECO-LOGICALMENTE," it said. His wide-angle photo was better than
my close-up shot because he'd framed the sign at dusk, in the broader
context of the cityscape.

I felt a sudden kinship with this man, Mik Awake, who'd noticed the
same thing I had, from a broader perspective. The two signs were clearly
of a piece. Intrigued, I did some Internet sleuthing, just as I had with the
Know Your Rights murals, and discovered that they were part of a larger
series by environmental artist Justin Brice Guariglia, in partnership with
the three-year-old Climate Museum and the mayor's office. Here was
yet another public art project that might offer instruction, or hope, if I
could follow the pattern. All in all, there were ten climate signs staged in

public parks across the city's five boroughs—many of them in low-lying neighborhoods near the water, most vulnerable to flooding.

The Climate Museum's website described Guariglia's project as an effort to confront New Yorkers with how global warming affects our city now, to "break the climate silence and encourage thought, dialogue and action to address the greatest challenge of our time." To that end, the signs' messages were programmed with translations in languages spoken in the various neighborhoods in which they appeared: Spanish, French, Russian, and Chinese. Thus embedded in the diverse cultural landscape of New York City, the billboards projected a forecast of what we stand to lose with the rising sea. The Climate Museum's website also offered a city map indicating the locations of the ten signs, with clear directions via public transportation to each. Finally, it introduced an adventure: any-one who could prove they'd visited all ten signs would receive a prize.

I studied the map with a mix of obligation and gratitude. It reminded me of my earlier failure to process reality at the American Museum of Natural History. Here was an opportunity, perhaps, to do better—if not with my kids yet, then at least with another concerned citizen. I decided to take the artist's invitation to heart. On a lark, I asked Mik Awake if he'd be willing to navigate Guariglia's climate signs with me. Would he be willing to pivot his attention to this concern and join his attention with mine? Amazingly, given the time commitment, the soberness of the topic, the complicated semiotics, and the distance over which the pilgrimage would carry us, he said yes.

We had until the midterm election day, when the signs were scheduled to come down. By visiting one or two a week, between September and November of this unseasonably warm fall, we managed to witness them all. I've come to think of this period of my life—part scavenger hunt, part Stations of the Cross—as "Thursdays with Mik." By now we are no longer strangers but friends. What follows is an account of our journey to grasp the effects of global warming on the place where we live.

The first sign Mik and I saw together sat at the end of Pier 84 in Hudson River Park, halfway between his neighborhood and mine, in what was commonly known as Hell's Kitchen. Thanks to real estate development, it's now often referred to by the tonier names of "Clinton" and "Midtown West." To get there, I took the downtown A to Forty-Second

Street and pushed west through the crush of tourists gazing up at the digital billboards of Times Square, wondering if those poor suckers knew they were looking at the wrong signs.

Only someone not from New York City would describe Times Square as the heart of the metropolis. Most of us who are native to the city steer clear of it, especially on New Year's Eve. But today, it couldn't be avoided. Passing through the clogged commercial district on my mission, I recalled an unnerving image from the short film *two°C* by the French filmmaking duo Ménilmonde. New York City is depicted in this movie as an Atlantis in the making, subsumed by the rising waters, with the Hudson and East Rivers converging to swallow a Times Square devoid of anyone to watch the blinking ads.

Is it possible to be haunted by the future as well as the past? The precise and intimate term for this feeling is "solastalgia," the desolation caused by an assault on the beloved place where one resides; a feeling of dislocation one gets at home. I suppose one might feel this in the case of war, domestic abuse, or dementia. I know I felt something like it after 9/11, back when I lived in Brooklyn with the alcoholic. But the difference with environmental upheaval is the ingredient of guilt. I walked past the bright theater marquees and the slovenly Port Authority bus station, the brownstone on Forty-Third Street where my friend Cathy had thrown my baby shower in her rent-controlled garden apartment, at which my mother had presented me with the first baby's quilt. I walked past the high-rise on Tenth Avenue where I'd screwed my high-school boyfriend on his parents' ratty foldout couch—past my former selves, and the ghosts of twentieth-century peep shows and nineteenth-century slaughterhouses, to join my companion at the waterfront.

Here is Mikael Awake, a few weeks shy of his thirty-seventh birthday. He's the kind of guy who's at ease quoting Paul Éluard ("*La terre est bleue comme une orange*") and whose friends ask him to officiate their weddings—a contemplative, caring, stylish man; the hardworking son of Ethiopian immigrants. Mik's last name is pronounced "a-wə-kə," but its meaning in English fits his character. That is, the brother is politically "woke," stumping hard that election cycle for Stacey Abrams to become the nation's first Black female governor down in Georgia, his home state. Mik grew up as one of few kids of color in the schools of Marietta. When

10.) *Climate Denial Kills*, Pier 84, Hudson River Park, Manhattan,
photographed September 27, 2018

I asked him what that was near, he joked, "Racism," before conceding,
"Atlanta."

According to the logic of social media that has shrunk the planet,
we're separated by only one degree, sharing several acquaintances and
interests in common. Both of us are writers and teach writing at the City
University of New York—a vocation, we agree, that brings us closer to
the city by putting us in touch with its strivers. We might eventually have
crossed paths another way. He'd been following me on Twitter. But it was
the climate signs that brought us together.

My instinct is to focus my lens on Mik, but, unwittingly, he's already
taught me to take a step back. Still, the viewfinder can't capture it all.
What I mean to show is too big. I settle for framing the tension among the
human, the landscape, and the sign. This method will become our pattern.

To the right of the frame floats the former aircraft carrier USS *Intrepid*;
to the left lies the Circle Line cruise ship departure site. The Hudson rolls
by behind Mik. The Lenape called it Shatemuc, meaning, the river that
runs both ways. Beneath the river, the busy Lincoln Tunnel connects to

New Jersey on the other side. Flooding of the tunnels into and out of Manhattan (along with flooding of the subway and energy infrastructure) were identified as serious vulnerabilities in *Climate Change and a Global City: Metropolitan East Coast Report*, a pre-Sandy document that examined global warming impacts in New York.

Cyclists zipped along the two-lane greenway in front of Mik—one of the many routes I rode before having kids. The bike path is but one perk of Hudson River Park, refurbished from the crumbling docks of a once grubby industrial waterfront as part of the city's renewal wherein, over the last generation, former warehouse districts have steadily transformed into luxury housing. Sandy wreaked nineteen billion dollars in damage to the city, yet the rate of development along our coastlines has only increased since that superstorm. We have more residents living in high-risk flood zones than any other city in the country, including Miami. Local city-planning experts are rightly concerned about how we'll cope when the next great storm surge inevitably strikes.

After Sandy, Mayor Bloomberg declared that New Yorkers wouldn't abandon our waterfront. He, and Mayor de Blasio after him, worked to revise codes to make new buildings more climate-resilient with flood-proofing measures such as placing mechanical equipment on higher floors, but each mayor has seen real estate interests in waterfront development as too precious a political constituency to suppress. In my picture, Mik faces Twelfth Avenue, across which, and slightly to the southeast, construction on the glassy towers of the huge Hudson Yards project is topping out, surrounded by a phalanx of cranes. Hudson Yards is one of many recent projects sited within the NYC climate change panel's projected floodplain for 2050, when sea level rise could reach 2.5 feet—meaning that all the tall buildings Mik and I behold cropping up at the waterline are shortsighted in that, gradually, they'll find their foundations inundated.

Investors tend to think in the short term—in the length of mortgages, according to the timelines of insurance policies. Even the newspaper headlines are guilty of this fallacy, tending to present Armageddon in the near future. Whether we frame it as twelve years away, or thirty, or fifty, or refer to ourselves as the "last generation that can stop climate change," we seem to keep pushing back the clock, as if the countdown to the ball drop has only just begun. But when did the clock start ticking? We wondered, Mik and I, if what we were experiencing now was the aftershocks of conquest, colonialism, and chattel slavery.

Over Mik's shoulder, the sign flashes its warning. How well the city's personality coheres in the bustling elements that surround my new friend: defense, tourism, transport, development, expansion, confrontation, bull-headedness, and art. Yet my picture, like that short disaster film, is eerily pastoral, as if Mik were the sole survivor of nascent catastrophe.

His face registers concern. I won't venture a mutual diagnosis of eco-anxiety, the emerging condition described by the American Psychological Association as the dread that attends "watching the slow and seemingly irrevocable impacts of climate change unfold, and worrying about the future for oneself, children, and later generations." I'm no clinician, and such a prognosis pales in comparison to the plight of a person in Puerto Rico still without power post-Maria, or a farmer in India driven to suicide by his perennially scorched crops. We're privileged not to have been directly hit yet, as so many have been in the Global South. It's our temporary luxury to consider global warming intellectually rather than materially. We have each in our own way, as Black Americans, more immediate threats to battle. Nevertheless, Mik confides in me that his stress makes him compulsively pick at his face. I confess to habitually clenching my shoulders, and to the nerve pain resulting from the Atlas-like tilt in my neck. The stimuli are not exaggerated. The glaciers are melting. The water is rising. These are our bodies' signs.

The second and third signs we toured were located on Governors Island. Situated in New York Harbor between lower Manhattan and Brooklyn, the island was used from 1755 to 1996 as a military post by the US Army and Coast Guard. Dredged debris from the harbor and landfill from subway excavation were used in the early 1900s to more than double the island's size—an illustration of the innovative if hubristic remaking of topography that's characterized the city for centuries. In 2003, the city bought Governors Island from the federal government for a dollar. Since its transfer to the people of New York, the island has been used as a public park, accessible by ferry to day-tripping civilians like us.

There's a public high school on this island. An organic farm. Artists' studios. Playgrounds. A fifty-seven-foot slide. A "glamping" (glamorous camping) retreat, costing at minimum five hundred dollars a night. There were even, at one point, a pair of friendly goats named Rice and Beans, and a miniature golf course. Its elevated hills were designed with climate change in mind. The cinematic view it offers of lower Manhattan is glorious—the stuff of a Gershwin score. I had never been to Governors Island before, though it's a place my children would fancy. The October day when Mik and I made our visit was warm enough for shorts.

9.) *Human Agenda Ahead*, Castle Williams, Governors Island, photographed October 4, 2018

To the left of my picture's frame stands the circular sandstone fort, Castle Williams, built in 1811 to defend the rich Port of New York against a naval assault that never came. Mik looks past the climate sign toward the dense cluster of buildings of the Financial District, at the ghost limbs of the Twin Towers and the blue spike of the so-called Freedom Tower at One World Trade Center sticking up like a middle finger. That precious spit of land beneath Mik's gaze seats a five-hundred-billion-dollar business sector that influences the world economy. Of course, the Financial

District wasn't the only part of lower Manhattan devastated by Sandy. Within a ten-mile perimeter, ninety-five thousand low-income, disabled, and elderly residents suffered because all of downtown—like Governors Island itself—lies in a floodplain. In downtown Manhattan, as in other low-lying neighborhoods at the water's edge, communication and transportation were cut off, infrastructure was damaged or wrecked, and thousands were without running water and power. Mik and I contemplate the irony of the prior use of Governors Island for defense when the real threat of attack to the city is the water surrounding it.

Also under Mik's gaze is a tidal gauge off the southern tip of Manhattan. Measurements taken there indicate roughly a foot of sea level rise in the last century. Climatologist Cynthia Rosenzweig, who served on the IPCC, has compared sea level rise to a staircase. "The twelve-inch increase in NY Harbor over the last century means we've already gone up one step. When a coastal storm occurs, the surge caused by the storm's winds already has a step up. Continuing to climb the staircase of sea level rise means we'll see [a] greater extent and frequency of coastal flooding from storms, even if the storms don't get any stronger, which they are projected to do," she said in a post-Sandy interview for Climate.gov.

So what are city planners doing to protect lower Manhattan? In 2013, in Sandy's wake, the Department of Housing and Urban Development's Rebuild by Design competition invited proposals for climate-resilient flooding solutions to protect against future storms. One of the seven finalist designs, Bjarke Ingels's BIG U, a vast ten-mile barrier system with a series of levees, was set to begin its first phase of construction in the spring.

Naysayers argue that such barriers won't save Howard Beach and other parts of southern Queens and Brooklyn; that a seawall enclosing the narrows of New York Harbor still leaves Long Island in trouble, that these halting projects to save the city will take too long, or fall short of the watermark, shifting the burden to stay afloat onto our children. Meanwhile, students at the high school on Governors Island are gamely helping to restore the ecosystem of the harbor with oysters, which, before they were killed off with the dumping of toxic waste and raw sewage into their reefs, once served as the base that made this estuary one of the most dynamic, biologically productive, and diverse habitats on the planet. It occurs to me that this hopeful act contradicts the apparently anthropo-centric sign, "HUMAN AGENDA AHEAD," insofar as other forms of life are being considered.

"What do you think this sign means?" Mik asked me. It was the most enigmatic of Guariglia's messages. The signal seemed to suggest both itself and its opposite—accusing the emblem of capitalism in the back-ground of a greedy agenda that's laid waste to the planet while perhaps, at the same time, appealing to its beholder to be more humane, putting people before profit.

Later, back at home, I called the artist at his Brooklyn studio, while my kids played The Floor Is Lava, to ask about the climate signals, and to clar-ify this one in particular. "That one came out of the notion that we live in a corporatocracy," Guariglia explained, quoting the bon mot that it's sim-pler to imagine the end of the world than the end of capitalism. "Corpora-tions want this news hidden. Big business has had next to no incentives to care for the earth, though you'd think the preservation of the human race would be incentive enough." The project was meant to compel people to think ecologically, he told me, and was influenced by the object-oriented ontological writing of eco-philosopher Timothy Morton, as well as the text-based work of neo-conceptual artist Jenny Holzer:

> We need to get more human-focused but not more anthropo-centric. As a human, I'm on the same ontological playing field as garbage and galaxies—not above. As a visual artist, I'm trying to engage with a huge thing operating on several levels, requiring sev-eral languages. This is an exceptionally urgent problem. It needs to get out into the broad public and raise consciousness. That's the responsibility of artists and writers—not the corporations.

I know from a photograph that Guariglia's left arm is tattooed with a wavy line representing four hundred thousand years of carbon dioxide levels in Earth's atmosphere. It spikes at his pulse point and encircles his wrist like a handcuff. He embodies this stuff. I wonder if making his art keeps him from physical pain, as making pilgrimages has done for me. Why the road signs? I asked.

"The medium of the highway sign is embedded with so much information," he said. "It's hyper-accessible and a symbol of authority. People see roadwork signs and think, *Government*. They have a limbic response. I wanted to communicate on a broad public level and have people subconsciously absorb it."

I confessed to having difficulty absorbing the message, and to feeling very small in the shadow of the signs. Guariglia understood.

Because he'd recently traveled over Greenland with NASA to photograph quickly melting polar ice caps, he spoke on a more personal level about a particular pockmarked "galloping glacier" (so called for the rapidity of its flow): "Could my brain really make sense of an object warehousing thirty-eight million Olympic-sized swimming pools of water? Or the scale of a hundred-thousand-year-old hunk of ice the size of California on the verge of calving from an ice sheet? No, but we have to talk about what happens when all this ice melts. *Where's it going?*"

8.) *Injustice*, Yankee Pier, Governors Island,
photographed October 4, 2018

Mik asked me at Yankee Pier, on the other side of Governors Island, what I thought of the signs as art. I considered whether or not I liked them. It was as if, after a doctor diagnosed one of my kids with cancer, someone else in the waiting room of the ER asked my opinion of that doctor's beard. The signs didn't offer any solace. I hated what they suggested. I did like that they'd brought us to a place offering fresh angles on the city. Mainly, I liked the signs for connecting me with Mik. It is not so easy to make new friends in middle age. Having children in common with the other parents at the playground is seldom enough cause for deep connection.

When I asked Mik, in turn, whether he liked the signs, he answered that they made him feel less alone.

Before leaving the island, we stopped at the Climate Museum's temporary hub in the Admiral's house, within a cluster of stately buildings that once served as officers' quarters. There, visitors were invited to simulate their own climate signs. These read like postcards from the brink, bumper stickers, or protest slogans:

SEE YOU ABOVE SEA LEVEL!
ISLAND NATION JUSTICE.
MAKE AMERICA GREEN AGAIN.

A pamphlet informed us that Earth's glaciers hold enough water to raise sea levels by roughly 230 feet. Mik wrote:

WATER IS THE IMMIGRANT THEY SHOULD FEAR.

The Climate Museum is the nation's first museum dedicated to climate change. I later asked its director, Miranda Massie, to comment more elaborately about the organization's mission. Massie responded in an email about how art, even art with a bleak message, can inspire us, "consciously or not," by reminding us of human creativity and potential:

> Museums have strengths including popularity and trust that make
> them essential to the cultural shift we need to see on climate. Art in
> particular has the power to move the needle on climate engagement
> and action for a couple of reasons. It reaches us physically, emo-
> tionally, and communally, where we really live. Climate change can
> seem abstract even in the middle of a hurricane. To fully confront
> the immense challenge posed by the crisis, we need a visceral sense

of reality. We also need to be able to experience the full range of
emotions art evokes—awe, grief, surprise, fury, tenderness. . . . And
perhaps most of all, we won't make progress without each other—
and art builds community. It is a soft pathway into climate dialogue
that lifts climate out of perceived polarization and stigma, allowing
us to break the climate silence.

   To talk about this subject seriously is to risk being called a doomsayer
or a scold. Massie told me that 65 percent of people in the United States
are worried about climate change, but only 5 percent of us speak about
it with any regularity. (Since our correspondence, the number of Americans fearful of climate change has been identified by a Yale poll as rising
to more than seven in ten.) In truth, Mik and I talked less in the beginning about climate change than about our lives—his book in progress, our
respective lesson plans, the imminent midterm elections, the costumes
my children were planning for Halloween . . . But now I know we were
also stumbling together on a path toward new language, led by the signs.

7.) *Caution*, Flushing Meadows Corona Park, Queens,
photographed October 11, 2018

The following week, we ventured out to Flushing Meadows Corona Park, in Queens, to see the fourth sign. The park was empty that Thursday on account of a torrential downpour. We were feeling the sultry edges of Michael, a Category 4 hurricane that had made landfall in Florida the day before. The radial pathways leading to the Unisphere were washed out. In the nearby isthmus neighborhood of Willets Point, where permanently flooded streets are bordered by Flushing Bay and the Flushing River, hundreds of small businesses are being torn down and relocated to make room for a multibillion-dollar megaproject. By the time we reached the sculpture, notwithstanding our umbrellas, Mik and I were both soaking wet. Visibility was so poor through the sheets of rain that we couldn't immediately spot the sign, and so we orbited Earth with no atlas, like two lost satellites, until we found it by Africa, stuck on the word, "CAUTION."

On the new Google Earth plug-in Surging Seas: Extreme Scenario 2100, created by Climate Central, which uses data from a report by the National Oceanic and Atmospheric Administration (NOAA) to make projections based on the country's greenhouse-gas emissions, the future map of Queens shows that, unless the United States transitions to clean energy alternatives, Flushing Meadows Corona Park will be engulfed. So will JFK and LaGuardia airports, sections of northern Astoria, College Point, and most of the Rockaways, while large swaths of Long Island City will be submerged in water. You can zoom in tightly on the map. From an unscientific perspective, the flooding can appear biblical, like the wrath of a punishing God. For example, the site of New Park Pizza—the pizza joint in Howard Beach where, in 1986, a mob of white thugs wielding baseball bats reportedly taunted twenty-three-year-old Michael Griffith and his friends, "What are you doing in this neighborhood, niggers?" and then attacked and chased them toward the Belt Parkway, where Griffith was killed by an oncoming car—will be mercilessly drowned.

Mik stands on a step, to avoid a puddle. The 350 tons of the steel Unisphere, which was the central icon of the 1964–1965 World's Fair, is more imposing in person than I had anticipated. Yet, in the context of the sign, and softened by a veil of mist, the Earth it represents appears fragile, nearly delicate. The gauzy material of Mik's shirt echoes the fountain spray. Under his umbrella, and next to the lamppost dividing my picture,

Mik looks like a version of Gene Kelly, too pensive to dance. The lamp at the top of the post marks South Sudan, just west of Ethiopia, whence Mik's parents fled Mengistu's Red Terror.

We discussed the comparative effects of climate change in the Horn of Africa and here on the eastern seaboard of the United States. While that part of the world is drying out, leading to shriveled crop yields and water shortages that in turn contribute to conflict, famine, and forced migration, this part of the world is getting more precipitation. Between 1958 and 2012, the Northeast saw an increase of more than 70 percent in the amount of rainfall measured during heavy precipitation events, a greater increase than in any other region in the nation, according to the EPA. That we're getting warmer and wetter here, however, doesn't mean we won't also feel drought conditions in summer. We will.

All-time heat records were set all over the world last summer, including in our global city. Some days were so hot I questioned whether it was safe to send my kids to day camp. By 2050, New York's average temperature is projected to rise between 4.1 and 6.6 degrees Fahrenheit, while annual precipitation is expected to increase between 4 percent and 13 percent. The regional trend toward more rain has exacerbated localized flooding—not only during big storms but also as a matter of course at high tide, since the ocean is already brimming. Street floods are a regular nuisance in some low-lying areas of Queens, like Hamilton Beach. Residents there have grown accustomed to swans and fish swimming in knee-high water in the middle of the road when the moon is full.

Perhaps the most common place the average New Yorker experiences flooding is the subway. Even on dry days, the MTA is tasked with flushing from its network some thirteen million gallons of groundwater with overworked pumps. When flash floods from heavy rain swamp the tunnels, those of us who commute are routinely delayed, spending more and more time griping underground.

Out in Queens, the 7 line is elevated. Mik and I didn't stay long, not just because of the inclement weather but also because he was scheduled that afternoon to sign on a town house he and his wife were purchasing in Flatbush, Brooklyn. They made sure the property lay outside the floodplain before making their offer. I made a mental note—should we ever buy a house in the city, we'd be wise to do the same. Sloshing our

way back to the train near the Mets' Citi Field ballpark, he showed me
the inscribed Montblanc pen he planned to sign the contract with—a
graduation gift from his brother. He looked the part of a proud first-time
homebuyer on the verge of the American Dream.

6.) *Human Agenda Ahead*, Sunset Park, Brooklyn,
photographed October 18, 2018

Until they make the move to their new home, Mik, his wife, and their
dog, Chester, will still live in the low-lying waterfront neighborhood of
Sunset Park, where nearly a third of the residents live below the poverty line. I joined him there in the park of the same name to visit the
fifth sign, the one he had tweeted back to me. Like other disadvantaged
communities, this one bears the burden of environmental pollution and
impacts of climate change. In addition to being susceptible to flooding,
Sunset Park endures poor air quality because of passing traffic on the
Gowanus Expressway and three nearby fossil-fuel-burning power plants
whose pollution, ironically, adds a pretty afterglow to the neighborhood's
already remarkable sunsets.

Mik and I met hours before sunset, at high noon. In my picture, he stands squarely on his shadow, almost camouflaged by the trunk of the tree that appears to branch from his head. The dark blue water of the upper bay is just visible at the horizon line. What you can't see is the gang of tough old Chinese women doing tai chi behind me by the tennis courts, in the sudden cold. They belong to a large Chinese community in Sunset Park, which boasts New York City's largest Chinatown. Appropriately, the sign addresses that part of the neighborhood's demographic in Chinese. Auto-correct as Freudian slip: when I later asked Kimm, the Chinese midwife who delivered my children, to translate the sign, she texted, "It's weird— 'Human Agenda for the Futility.'" And then, "Oops!—'Future.'"

Donald Trump declared climate change is a fabrication on the part of "the Chinese in order to make US manufacturing non-competitive." He also appointed fossil-fuel advocates to lead the EPA and Department of Energy. What makes this denial and disregard so egregious is that the United States is the second largest contributor of carbon dioxide to our planet's atmosphere, though we're home to only 4.4 percent of its pop-ulation. It would take four Earths to provide enough resources for all if everyone in the world consumed as much as we do in the United States.

Interestingly, Sunset Park is home to the city's "first grassroots-led, bottom-up climate adaptation and resiliency planning project." Following Sandy, community members organized into a block-by-block, building-by-building plan for action called the Sunset Park Climate Justice Center. The organization's mission includes supporting local leaders "to coor-dinate allocation of community resources and mitigate the impacts of future severe weather, including the possible release of harmful chem-icals."

Since many workers in Sunset Park are employed by natural gas and power plants that could be shut down or curtailed as new climate reg-ulations take effect, activists have appealed to the governor to let com-munities like Sunset Park lead the development of a transition plan for workers in the fossil-fuel industry to find well-paying new jobs in the regenerative-energy economy—in other words, to play an active role in their own adaptation. One example of climate adaptation underway in Sunset Park is a new initiative to shift to renewable energy on a cooper-ative ownership model. An eighty-thousand-square-foot solar garden is under development nearby, on the roof of the decommissioned Brooklyn

Army Terminal. It is planned to start operating this year, open especially to low-income residents. Aiming to serve some two hundred families as well as businesses, the solar array will be one of the country's first models of a cooperatively owned urban power supply, cutting energy costs and emissions for subscribers.

It's inspiring to uncover local action taking place despite federal inaction. It's a drop in an ocean-sized bucket of hypoxic, tepid water, but it's inspiring. Mik told me that he and his wife were looking into applying for a solar rebate initiative and exploring how to green the roof of their new house. I thought it would be impertinent to ask if they planned on having kids, though I believe he'd make a great dad. We discussed their plans for renovation while Chester chased his tail in widening circles.

5.) *Climate Change in Effect,* Beach Ninety-Second St., Rockaway Beach, Queens, photographed October 25, 2018

The sixth sign stood way out in the Rockaways, in Queens. I could have traveled there on the A train, but I chose to go by water. As usual, I was running late. I rushed through Battery Park's tight warren of streets over the African Burial Ground in time to catch the ferry at Wall Street/Pier 11. The Freedom Tower jutted above, like the blade of a sundial. I ran past jackhammering workmen in orange vests and hard hats, the stock exchange, a discount store that used to be an outpost of the Strand, and, on tiny Maiden Lane, the ghost of myself aged twenty-five, disoriented and breathing dust, having walked over the Brooklyn Bridge on Rosh Hashanah to stumble upon Ground Zero a week after 9/11. I jumped at the blare of a horn: a Moishe's moving truck rolled through a red light, nearly mowing me down. I was struck by how quickly the city remakes itself; these narrow, cobbled streets were made for horses. I raced waterward down Stone Street, aware of the ticking clock. Motherhood had shrunk my time. This journey had a hard stop; I had to be back to pick up the kids when aftercare was done. Flanking the terminal were the Brooklyn Bridge and a noisy helipad, barges, water taxis, seagulls. At Slip A, one dockworker playfully grabbed another from behind: "Yo, you got documentation? This is ICE; assume the position." A ticket to Rockaway Beach cost $2.75, the same as a subway fare.

The ferry windows were grubby; seeing out of them was like watching a dream sequence. The shapes at the edges of Brooklyn were hard to discern: construction cranes, IKEA, a windmill, rooftop water tanks like the hats of witches, shipping containers, men with fishing poles. Mik embarked at Sunset Park, somewhat shaken—the day before, a bomb had been delivered to the CNN building, close to his wife's workplace.

On the boat ride, Mik and I discussed living in a post–September 11 state of alert, often fearing for our loved ones' safety. Recently, my children had been evacuated from their school, where they regularly practice soft lockdown drills for active-shooter incidents. They seemed confused by this. Their father and I will have to have a talk with them about gun violence, too, on top of the talk about police violence. This is something beyond the quotidian mental exhaustion that comes from urban living: behavioral studies have found city dwellers pay more attention to dangers and opportunities but less attention in general. In spite of our sensory overload, or because of it, the climate signs have demanded our attention.

As we were ferried over the water lapping at Brooklyn's coast, through Gravesend Bay, past the Wonder Wheel at Coney Island and Brighton Beach toward Broad Channel, Mik and I imagined the heightened state of alert that must attend living on the coast. And yet, we envied our neighbors on the coastline their view. When you live, as my family does, in a mid-rise apartment building with a foreshortened view of the building across an alleyway of battered trash cans, it's easy to forget you're near water. New York's master builder Robert Moses made it harder for pedestrians to access the rivers, creeks, straits, lagoons, and bays surrounding New York City by hemming in the city with so many expressways. It takes a mental leap to make the worthwhile plunge.

Neither Mik nor I had ventured to Rockaway Beach before; we'd become tourists at the outer reaches of our own city. The ferry took its time. The sky opened out. We felt the primal allure of water, the drag of the vessel upon it. Soon enough, our heart rates slowed. Melville writes, at the opening of *Moby-Dick*, about gravitating toward the sea: "It is a way I have of driving off the spleen."

A feeling of danger gave way to a feeling of opportunity. We disembarked at the Rockaway Peninsula, an eleven-mile strip of two-family homes and public housing projects. A dome of gold light hovered over it, cast by the double reflection of Jamaica Bay and the Atlantic Ocean on the other side. On the free shuttle bus that drove us closer to our destination, I wondered aloud if people who live and work on the water feel happier. A local man with a Russian accent answered yes. "Why do you think people come all the way out here with their wetsuits and surfboards on the subway?" he asked. "To relax!" Then, he kindly directed us to Beach Ninety-Fourth Street. The beach was empty. Mik and I couldn't stop smiling—we felt like we were playing hooky. We surveyed the vast Atlantic and breathed.

The portrait I made of Mik at the waterline captures that repose. But to my eye, it also looks uncannily like the Andrew Wyeth painting *Christina's World*. The resemblance lies in the nuance of shadows and light, the waving grass, my subject's backward-facing posture, and the property on the horizon line (in this case, a Mitchell-Lama housing complex rather than a farmhouse and outbuildings) that appears imperiled by a looming, unseen force.

The sign sits on the new $341 million concrete boardwalk, built to replace the wooden promenade wrecked by Sandy. "CLIMATE CHANGE IN EFFECT," it warns, in Russian. The more resilient boardwalk has been built at a higher elevation, with sunken steel pilings above a retaining wall designed to keep sand from being pushed into the streets.

Rockaway was among the New York City communities hardest hit by Sandy. When the hurricane slammed into the unprotected peninsula, flooding it with a fifteen-foot storm surge, the old boardwalk was ripped from its moorings to smash into beachfront properties. A raging six-alarm fire took out some one hundred homes in the area of Breezy Point; another blaze decimated a commercial block, and power was down for weeks. The nearby Rockaway Wastewater Treatment Plant was inoperable for three days, during which hundreds of millions of gallons of raw sewage were released into the waterways—a gross example of why the city must retrofit its wastewater facilities for higher tide levels, to avoid drowning in its own shit.

Post-Sandy, the federal government replenished Rockaway Beach with enough sand to fill the Empire State Building twice. An Army Corps of Engineers project has proposed to build, on the beach side, eighteen-foot reinforced dunes and thirteen new jetties along the peninsula to help stockpile more sand, which would mitigate the force of storm tides, and on the bay side, a $3.6 billion system of levees, gates, and floodwalls to control the level of water. Some blue-collar residents along the Rockaway rail line across Jamaica Bay have used funds from the city's Build It Back program to raise their houses higher up on stilts. But the Rockaways were built on a sandbar that geologists have argued should never have been developed in the first place. A troubling amount of restocked sand on Rockaway Beach has already been eroded by the relentless action of the waves. Outside my picture's frame, the sun-dazzled ocean kisses up to the dunes at narrowed parts of the shore, caressing the beach up as far as the new boardwalk.

How to reconcile these twin feelings of pleasure in the city's enjoyments and terror of its threats? I have learned that the world is running out of sand; that every second we're adding four Hiroshima bombs' worth of heat to the oceans. I think of an hourglass. I think of my kids. But for now, being hungry, Mik and I set out for lunch.

Visiting the seventh sign, Mik humored my kids by agreeing to wear a Halloween mask in St. Nicholas Park, where I first saw the sign. The ghostly white mask is from the production of an adaptation of *Macbeth*

4.) *Climate Change at Work*, St. Nicholas Park, West Harlem,
photographed October 29, 2018

called *Sleep No More*. My kids were in costume, too. I brought them along on our errand to reckon with the climate sign after their school's Halloween parade. My big boy was dressed as a killer scarecrow, and my little one, a ninja. I was their mama wolf.

After I shot Mik looking like a nightmare in the mask with no mouth, standing before the sign that originally caught my attention, he was sweet enough to borrow my iPhone to make a family portrait of us. The kids were oblivious to the sign, using the park for its highest purpose—play. Their games had changed since the days when we took them to J. Hood Wright Park, but now, as then, they could not keep still. Outside of the frame, they collected acorns, greeted dogs, poked sticks in the mud, and tumbled joyfully down the hill.

Not far from here in West Harlem is Riverbank State Park, where their father and I take the kids ice-skating in winter—where I was coming from when I saw my spark bird, the pair of burrowing owls. Riverbank State Park is spread alongside the Hudson River above the North River Wastewater Treatment Plant, where the pumps are receiving an upgrade as part of a renovation to improve climate resilience. From the skating rink, you can see the plant's exhaust stacks rising like a pack of tall white cigarettes. When the wind is strong, it smells of rotten eggs. Recent air samples from the site showed levels of formaldehyde that exceeded guidelines established by the city's Department of Environmental Protection, raising the ire of environmental justice groups fed up with the chronic placement of toxic and hazardous waste sites in low-income Black and brown neighborhoods.

Formaldehyde, exposure to which has been linked with cancer, can be created by incomplete combustion of methane gas produced during the process of wastewater treatment. Molecule for molecule, methane's a stronger greenhouse gas than carbon dioxide, though it's less abundant in the atmosphere. Its main human-derived sources are agriculture, such as cattle farming, waste (such as from landfills), and the fossil-fuel industry. "Really, the best thing you can do to save the planet is to kill yourself," joked a stand-up comedian I had seen at a club in SoHo with a girlfriend the night before. "The second-best thing is to stop eating beef." Earth has as much as six times more methane concentrated in the atmosphere than before homo sapiens emerged. With the thawing of the permafrost regions of the Arctic, more methane is being released. My kids are at the age of scatological humor where they love fart jokes, but as funny as I found that comedian, there's not much to laugh at in this.

The sign stands high above Mik, on the bluff toward Sugar Hill. Uptown, we're rolling in hills, some so prominent that they grant sweeping southern views of Manhattan, razed in the nineteenth century of its forests, farms, rocks, and slopes to create the rectangular grid. Back when my son was obsessed with violent weather and I told him we were safe in the Heights, he understood it was an evasion. We aren't disconnected from the other parts of the city, just as our country isn't disconnected from the other parts of the world.

In the silent white mask, Mik faces the entrance to the 135th Street subway at the bottom of the slope. The B stops here, and the C. The subways are the blood vessels of our body politic, just as nerves in the spinal cord connect the central nervous system. Even at three, my boy understood this. It's hard to imagine what the city would be like without its trains, should flooding disable the system altogether; what the spire of the Empire State building, in midtown, would look like from the rump of Sugar Hill—a buoy? I can't help wondering what the city will look like to our kids when we're no longer here.

At night, when I tuck in the little one, we play a game. I ask if he knows how much I love him, and he replies in the following ways: *Infinity hundred infinity, more than all the world's worlds. Long after you're dead, and I'm dead, and the world is dead, too.* To each reply, I answer, *More.* Maybe motherlove is a hyperobject, but so, Guariglia has told me, is climate change. For all the ferocity of my love, I'm powerless to protect my kids from the mass extinction we're in the midst of that could eliminate 30 to 50 percent of all living species by the middle of the twenty-first century. Why is this not the core of the core curriculum? Why aren't we all speaking about this? Jonas Salk's question: Are we being good ancestors?

3.) *No Icebergs Ahead*, St. Mary's Park, Mott Haven, Bronx, photographed October 31, 2018

Mik told me that the first user-generated question that came up in his search for St. Mary's Park in the Mott Haven section of the Bronx was about safety. The eighth sign, on a hilltop across the dog run, was defaced with graffiti. Two dicks spouting cum and the dictate "Eat my ass" competed with Guariglia's flashing warning signals: "FOSSIL FUEL-ING INEQUALITY," "CLIMATE DENIAL KILLS."

"FUCK YOU!" someone had also scrawled on the sign's orange base, which was surrounded by trash. Maybe teenagers had done it, like the two girls passing behind Mik in my photo. I like to think they wouldn't have, had the art offered comfort or beauty, or had the sign been useful in a practical sense. But in the context of the southeast Bronx, where failing infrastructure is a fact of life, the graffiti read to Mik and me like an act of protest against the hazard sign itself.

It would have taken me four trains to get here from upper Manhattan, though it isn't far as the crow flies, except that the third train wasn't running to carry me to the fourth, and so I got off at Yankee Stadium and walked three miles to reach St. Mary's Park—past the Bronx County Courthouse, through the Melrose Houses, and behind a trans woman in fabulous leggings printed with silver skulls, who clutched at her tooth-ache on her way to a storefront dental clinic, moaning, "Why is the demon bothering me, why?"

Last week, single-use plastics were banned by the European Parliament, while the news came that the sea is absorbing far more heat than we'd realized, and that it's too late to save the earth by planting trees. Even if we covered the planet with trees, it would be too late. It is already broken. Another sign appeared in my feed, this one taped in the window of a bookshop in Cornwall, England: "Please note: The postapocalyptical fiction section has been moved to Current Affairs."

The densely packed Bronx is the poorest of New York's boroughs, making it more vulnerable to heat of every kind. It isn't easy for the people who live here to leave. What does it mean, in a neighborhood where many forces choke possibility and freedom of movement itself is restricted by transportation that fails on the regular, to be told that climate denial kills? When we suffer an untreated toothache, the pain is so immediate that we can think of nothing else.

At Hunts Point, two stops closer to the waterfront on the 6 train, Mik and I discovered after crossing beneath the busy Bruckner Expressway that the ninth sign, in Riverside Park, was blank. Its solar battery had been jacked, the Master Lock popped, the orange encasement flung open like

2.) *Electric Relaxation*, Hunts Point Riverside Park, Bronx,
photographed October 31, 2018

the lid of a pilfered jewelry box. We were impressed by the thief's enter-prise in recognizing the battery's value. When I shot Mik sitting beneath the blank screen, I felt curiously relieved by the sign's defacement, its silence. By this time, at the ninth site, I had memorized the looping mes-sages of the hazard signs, just as I'd memorized the looping messages of the Know Your Rights murals, and could focus instead on the graceful posture of my friend, who'd opened his coat to enjoy the sun; the sludge of the Bronx River behind him; and the projects on the opposite shore.

Riverside Park is a little park, covering just 1.4 acres, a Faulknerian "postage stamp of native soil." Once a vacant lot and illegal dumping site, it's the first of a planned string of parks to be linked by a bike route as part of the developing South Bronx Greenway. Behind Mik is a fishing pier and canoe launch. I've been here before with my kids, to ride rowboats in the dog days of summer when folks inflate bouncy castles for the kids, hold all-day barbecues, and pump nineties hip-hop from big sound sys-tems, escaping the heat by the waterside. Although we're now deep into the fall and the only ones here, today is freakishly warm at 67 degrees.

Extreme heat is a dangerous element of our changing climate. It kills more Americans per year than any other weather-related event. Cities are often two to eight degrees Fahrenheit warmer than the areas that surround them because of the urban heat-island effect. At night, when our bodies need to recuperate from stress and heat, the contrast in temperature is more extreme, ranging as much as twenty-two degrees. This is due to urban design—our tall buildings, black rooftops, and dark pavement, which attract and absorb the sun's rays.

In an Arizona State University study of the summer of 2017, people of color were found to live in a census tract with more extreme urban heat than white people in nearly every major US city. A prior Berkeley study indicated that people of color are up to 52 percent more likely to live in urban heat islands than white people. We face more health risks during heat waves as a result. What's more, long histories of under-investment in Black and brown urban neighborhoods have made our communities heat up more and faster overall than the already hot cities we live in. Hunts Point has one of the city's highest rates of heat-related fatalities; 98 percent of its residents are people of color. The NYC Environmental Justice Alliance has made addressing the urban heat-island effect its top priority for climate justice because, while hurricanes and storm surges may smack us every five or more years, we can count on extreme heat to clobber us every single summer.

In the summer of 2018, 729 New Yorkers went to the ER for heat-related illnesses. New York City suffers a yearly average of more than 100 deaths attributed to extreme heat, but the Environmental Justice Alliance estimates the actual death toll to be a lot higher—depending on the variable criteria different researchers use to link deaths to heat waves, the annual number could be more than 600. A recent Columbia University study projected that by 2080, up to 3,300 New Yorkers per year could die due to heat-related causes exacerbated by climate change. In fact, New York is one of the most vulnerable cities in the developed world to the threat of urban heat.

The Hunts Point peninsula is an example of a hot spot within a heat island. It's a heavily industrial neighborhood where residents—who make, on average, twenty-two thousand dollars a year—often live in packed apartment buildings, with more heat-trapping surfaces and fewer green spaces to help cool the neighborhood. Who can afford an air conditioner on four hundred dollars a week, let alone the higher electric bill to run it, when there is rent to be paid?

The New Yorkers most at risk of heat emergencies live in fence-line communities of color like this one, burdened by decades of polluting infrastructure as well as poverty. For a long time, Hunts Point had one of the lowest parks-to-people ratios in the city, but over the past decade, locals have lobbied for more green space along its riverfront. To the left of my frame is a salvage yard heaving with crushed cars and scrap metal. At my back, across some railroad tracks, an idling produce truck is unloading cartons of ripe pineapples and strawberries. Hunts Point works as both mouth and ass for the city, getting food in, taking waste out: this neighborhood of thirteen thousand people has the world's largest wholesale produce market and over a dozen waste transfer stations. Along with the rest of the South Bronx, it handles nearly a third of the city's solid waste—meaning, there are trucks everywhere, all the time, piping out hot exhaust, heightening the risk of asthma attacks, and further broiling the air.

The city's Office of Recovery and Resiliency launched a "Cool Neighborhoods Plan," in 2017, to partner with grassroots organizations in efforts to mitigate the health risks of extreme heat in vulnerable communities. Strategies include painting surfaces white, planting trees, checking on older people who might be housebound in stifling apartments, greening roofs, and getting out the word about cooling centers—public air-conditioned facilities like libraries, community centers, and senior centers where folks can cool off for free. City officials piloted the initiative in three neighborhoods last summer, including Hunts Point. Without whole-cloth citywide initiatives toward cleaner energy, hot spots like this could be feeling summer days that are 15 percent hotter than today by the 2080s. Can you even imagine a 120-degree day? In the cities of New Delhi, Baghdad, Khartoum, Mexicali, and Phoenix, they're already here.

The unspoken threat remains in the frame, as does the tension among the sign, the human, and the landscape, but all the same, it was another beautiful afternoon for walking the city. Later on, I took my children trick-or-treating without need of their jackets. I felt uneasy, remembering how comparatively cool it was on Halloween when I was a kid, but also happy to witness their fun.

1.) *Vote Eco-Logically*, Snug Harbor Cultural Center and Botanical Garden, Staten Island, photographed November 5, 2018

The following week, Mik and I visited the tenth and last sign. We rode the Staten Island Ferry to get us closer to Snug Harbor, where the sign was stationed. It was the day before the midterm elections and gently raining. The air held an elegiac mist with an electrical charge: leaning over the ferry's wet rail, it was impossible not to conjure Walt Whitman thinking about us "ever so many generations hence."

"Whitman was one of my gateway drugs to literature, and these days I have such a hard time connecting to his voice, optimism, and vision of America," Mik admitted. "How hard it is now to have faith that future generations will even exist." We regarded the Statue of Liberty in a shawl of fog, and Ellis Island.

Climate change has become an often-unspoken contributing factor driving recent waves of immigration, such as the Central American migrant caravan used by the right-wing media to stir up ire as it neared the Mexican border. Mik grew quiet looking out at the white wake of passing ships: patrol boats, skimmers, tugs, barges, the ghosts of whalers and steamers, and the ocean liner *Queen Mary 2*. I asked what he was thinking about.

"The wildfires in California," he said, and all those fleeing fire, "but also how many more of us will be climate refugees in our own lifetime, all because folk turned greed into an economic system a few hundred years ago."

Belying his justifiable pessimism was the fact that Mik had just returned from Georgia, where he'd volunteered to drive elders without transportation to the polls for early voting. Understandably, he was nervous about the direction our country would go. There was a larger national anxiety about the feasibility of a "blue wave"—a Democratic sweep to win back the House. "I don't believe it," then president Trump said of his own administration's November report, which stated, "Climate change is transforming where and how we live." In the Republican stronghold of Staten Island, a borough known for voting against its own environmental interests, it cheered us somewhat to observe Democratic congressional election signs staked in the front yards of North Shore houses on the bus route to Snug Harbor.

Snug Harbor has been referred to as Staten Island's "crown jewel." People get married and shoot films there, but not the grimy seventies ones. Once a retirement community for aged merchant sailors, it's now a National Historic Landmark District set inside an eighty-three-acre park that runs along the Kill Van Kull tidal strait. Mik and I wandered among lush gardens freakishly still in bloom, a duck pond, a fountain memorial to local rescue workers who lost their lives on September 11, a farm, a surreal field of brightly colored lanterns including the giant head of a dragon, and several Greek Revival, Beaux Arts, and Italianate buildings—a chapel, a foundry, a theater hall, a hospital—remnants of the nineteenth-century seafaring community, repurposed for the arts.

Inside one of these stately buildings, now operating as a museum, we found Gus, a maintenance worker built like a fire hydrant, who bragged about Snug Harbor with charming civic pride. He was grateful to Jackie Onassis for her part in preserving the site and was glad to share its history with other New Yorkers—who, he admitted, tend not to visit the borough because it's difficult to access, and because of lingering bad PR over the Fresh Kills dump that grew, over the second half of the twentieth century, into the biggest manmade structure in the world.

Mik asked if the maintenance crew was doing anything particular to protect the landmarked buildings from sea level rise. Not that he was aware of, Gus said. "Like most New Yorkers, we only think about that stuff when it's too late."

On the subway map, Staten Island looks small and somewhat neglected, an afterthought boxed in the lower left-hand corner like a leftover chicken nugget. In reality, it's huge—over two and a half times the landmass of Manhattan, nearly as sprawling as the borough of Brooklyn. Situated in the crook of the New York Bight, the island suffered more than half of the city's forty-three deaths during Sandy, bearing the brunt of a storm tide that peaked here at sixteen feet.

Gus talked about the southeast shore of Staten Island, specifically the wooden bungalow community of Oakwood Beach, hit by the super- storm's worst flooding in a funnel effect that left some clinging to their rooftops while others drowned in their basements, trying to fix their sump pumps. That part of Staten Island is now rewilding with grasses, flowers, insects, possums, deer—returning to a natural wetland state. Most of the homes there were demolished after property owners took buyouts from the state government, choosing to relocate rather than attempt to rebuild—one of few examples in the city of managed retreat. As our coastlines become increasingly unlivable, this kind of deliberate migration away from the water's edge may grow more commonplace. In the meantime, a breakwater project is underway around the South Shore to protect those stalwarts who remain, seeding oyster beds to prevent coastal erosion, absorb wave energy, and clean the water, while the Army Corps is planning a $580 million seawall that many scientists claim would ultimately fail.

"Humans are stupid." Gus shrugged. "We want what we want, even when it don't make sense." He urged us to return to Snug Harbor for the winter lantern festival. He'd mistaken us for a couple. In a platonic sense, I guess we were. Ours was a marriage of inconvenient truth: I couldn't show a face of fear to my family, not while my kids were in the room, and since our apartment was so small, they were always in the room. So I showed it instead to Mik, who held my gaze without judgment and looked right back. Even though I knew I'd see him again, I felt a little sad that our assignment was ending. I enjoyed his company, and our journey gave ritual structure to my busy life as a working mother. The rain was growing heavier, and so we took cover under a gazebo.

In our final portrait, Mik took his hair down and folded his arms protectively over his middle. He looked older to me than he did in September, when we met by the first sign. "*La terre est bleue comme une orange.*" The world is blue like an orange. Orange is a color that strikes you in the gut. Van Gogh said it was the color of insanity. I am struck by the orange dominating this picture, turning it nostalgic: the traffic cone, the fall leaves, the gazebo walls, the base of the road sign, the text, and even the undertone of Mik's skin appear orange beneath the thunderclouds. The warm orange of autumn integrated with the bright orange of hazard.

"VOTE ECO-LOGICALLY," warns the sign, approaching its target.

The next day, a lot of us did just that. Bronx-born Alexandria Ocasio-Cortez, the democratic socialist who was voted to represent New York's fourteenth congressional district, had just posed the Green New Deal, which aspires to cut US carbon emissions soon enough to attain the Paris Agreement's most ambitious goal: preventing the world from warming any more than 1.5 degrees Celsius by 2100. "This is going to be the New Deal, the Great Society, the moon shot, the civil-rights movement of our generation," the young congresswoman said at a town hall meeting a month after the election, drawing on the cocky attitude of American exceptionalism. Who was I to say "too little, too late," when she's supplied me with the script to motivate my children with this rallying cry: "The only way we're going to get out of this situation is by choosing to be courageous."

I heard the same conviction in Justin Guariglia's voice when I interviewed him about his art. He described what it felt like to photograph the Arctic meltdown from the troposphere with a mixture of emotions: fragility, urgency, anxiety, awe—the simultaneous contraction and expansion of self that Whitman sang about. "You sense your own insignificance and the sublime," was how Guariglia put it.

My route with Mik across New York felt like a less glorious, though still noble, version of that crossing. I hold in my mind a new map of the city as a vulnerable and precious entity, both larger and smaller than I had understood; an appreciation of the water that binds us; gratitude for the prize of a friend I didn't have before (worth infinitely more than the tote bag bestowed by the Climate Museum—though that's cute, too); and the sinking realization that, eventually, we may have to migrate. Finally, I learned an altered sense of time, which I'll describe by paraphrasing the philosopher whose work inspired the climate signs that led us down a soft pathway out of silence into speech: we must awaken from the dream that the world is about to end; action depends on our awakening. When did the countdown begin? Let us reconsider the clock. Morton theorizes that the world has already ended. The ball dropped in 1784, he writes, with the advent of the steam train and the resulting soot that indelibly marked our footprint on the earth's geology during our swift carriage into the Industrial Revolution.

Riding the A train recently, we saw a showtime crew breakdancing to a remix of "It's After the End of the World, Don't You Know That Yet?" by Sun Ra and His Intergalactic Research Arkestra. Between 59th Street and 125th, the dancers took turns defying gravity, using the poles, hand-rails, ceiling, and floor to step, freeze, uprock, windmill, headspin, and air flare without kicking a single passenger in that rocking train car, before passing the hat. *Are they superheroes?* my second-born asked, mimicking their moves. My kid's a good dancer, in constant motion. He has a strong core. Since the age of three, he's been studying capoeira in the Bronx. This Brazilian martial art is dance, protest, music, self-defense, improvisation, beauty, play, and fight. *Yes, they are superheroes,* I told him, *and so are you.*

I think of the romance of the train, the iron horse that collapsed time and distance even as it began to undo us. How well my firstborn once loved trains. That boy is no longer here; in his place is another. In Mik's picture, my son stands as tall as my shoulder. Fittingly, for a kid whose latest passion is monsters, his favorite holiday is Halloween. At first, I begged him not to chew Mik's ear off about supervillains from the Marvel Universe and the darker actions of Greek gods. Then I stopped myself, since he receives too much negative attention already, and thanked my boy for his morbid curiosity. He is teaching us to pay attention.

The author with her children in their Halloween costumes, St. Nicholas Park, West Harlem,
photographed October 29, 2018

The signs hint at violence. I like how he and his brother are dressed in the context of the sign's bad news: "CAUTION." Now they are seven and nine. Their nicknames are Chief and Plum. Chief wields a scythe. Plum wears a katana tucked in his belt. One is stealthy; the other is fierce. My young look strong and alert. Good. They will have to be brave for the roadwork ahead.

2018–2019

PART II

PEOPLE LIKE HER

Vote Here, W. 138th St. and Hamilton Pl., West Harlem,
lead artist: LNY, Creative Art Works

Rufous Hummingbird, 3860 Broadway, Washington Heights, Manhattan,
muralist: Yumi Rodriguez

BIRDS SEE NO BORDERS

Yumi Rodriguez's grandfather used to call her colibri. That was his pet name for her: "hummingbird." *Vuela, vuela, colibri!* he would tell her, encouraging her to fly. She was raised by her grandparents on 161st Street and Broadway, across the street from the spot where she would later paint a mural in tribute to her grandfather. Like a lot of immigrants, Odalis Alvarez did a lot of things. He was a clockmaker, a blacksmith, a jewelry maker, a barber, an artist, and, eventually, a business owner; he owned a chain of barbershops in the hood. Yumi describes him as her support network—a quiet person, "a wise, cool dude" with whom she watched nature documentaries.

Yumi went on to earn a degree in animal nursing and now works as a veterinary technician to put herself through art school at Cooper Union, where she's pursuing public art. When Yumi's mentor, who does anatomical illustration at the American Museum of Natural History, learned she lived in Washington Heights, he told her about the Audubon Mural Project and put her in touch with Avi Gitler, the gallerist who spearheaded it. By this time, her beloved grandfather had grown ill. Yumi reached out to Avi to say she wished to paint a hummingbird—*un colibri*. He showed her the list of available species that hadn't yet been represented. That's how she chose the rufous hummingbird to paint on the riot gate of Romulo Barber Shop, the last place her grandfather worked. But really, she says, the bird chose her. Serendipitously, she'd read about the rufous hummingbird in a magazine that called it "the toughest bird on the block," with a furious heartbeat and one of the longest migratory journeys for a bird its size, from Mexico to Alaska, following a route of nectar-filled seasonal blooms.

"My grandfather always said I'd persevere," she says. "I share that characteristic with this little bird." She took care when designing her mural that it was "not just a pretty thing" but rather an image for people to spectate with a sense of respect. She included insects and plants, pollinators integral to the balance of the ecosystem to keep it from collapse. She also included this text: "*Un homenaje a mi abuelo, Odalis Alvarez, por su passion por la conservación de la naturaleza.*"

Yumi's grandfather never got to see the mural. He died of prolonged heart failure in the spring of 2020, when New York City's soundtrack was dominated by two remarkable strains of music: birdsong and siren.

≣

Avi Gitler's partners were interested in creating a dialogue about climate justice, he told me in a larger conversation about the aims of the Audubon Mural Project, but that was not his goal. Personally, he wanted to put up great art, intending to reach kids. "My dream is that twenty-five years from now, an ornithologist will reach out to me to say they grew up in the neighborhood and were turned on to birds by the murals."

Avi, a great-grandchild of Jewish refugees who came to Washington Heights from Europe, grew up in the neighborhood himself in the 1990s, when it was rough. His grandparents, born in the Heights, were kosher caterers. He attended the local Yeshiva University and its associated high school. With no formal background in the art world except for an education from the Met and the city's other great art museums, he started an art business that led to the establishment of his gallery, Gitler & ____, on Broadway in 2014. It was the only commercial art business in the neighborhood. He felt jazzed about participating in revitalizing the neighborhood, attracting patrons who would shop in local stores and eat in local restaurants. He chose the location in part for its proximity to Audubon Terrace, where, he says, "you can see world-class art for free at the Hispanic Society Museum—probably the only place where you could steal a Velázquez if you spent an hour thinking about how you were going to do it." He got permission from the neighboring store for an artist to paint a flamingo. Then a resident who worked at the Audubon Society noticed the bird and told Avi about the climate report. Soon after, the mural project began.

The project has the approval of the community board and partner-

Northern Parula, Blue-Headed Vireo, and Pine Siskin,
3898 Broadway, Washington Heights, Manhattan,
muralist: BlusterOne

ships with local schools, such as the Washington Heights Expeditionary
Learning School. (The fox sparrow was recently painted on an entrance
to the school, in honor of its dedicated environmental and climate science teacher, Dr. Jared Fox, who wants to develop a green corridor.) Several uptown artists were commissioned to contribute, including Yumi
Rodriguez, who painted the rufous hummingbird, and BlusterOne,
who lives on 151st Street. BlusterOne painted the American redstart,
for which they named a cocktail at the bar Harlem Public, followed by
Three Little Birds, in honor of Bob Marley, which Avi described as "a pos-
itive message in an age of carnage." Avi distinguished the project from
the Wynwood Walls in Miami, a public mural project directed by a real
estate entity to transform a blighted industrial warehouse district into
what the *Miami Herald* described as a lucrative "red-hot residential zone
with artistic soul." In Avi's view, the Audubon Mural Project is mindful
of the history of street art, "trying to give voice to the voiceless. Who has
more ultimate lack of agency than a population of birds?" He cited the
AIDS era and the work of Keith Haring as its origin, appearing "where
nobody wanted to go."

Gang of Warblers, 601 W. 162nd St., Washington Heights, Manhattan,
artist: George Boorujy

Now there are tours of the Harlem murals on Sunday mornings, offered through NYC Audubon, for thirty dollars a head.

Brooklyn-based studio artist George Boorujy, who painted *Gang of Warblers* on Dibond with outdoor house paints, wanted to depict them as a bunch of tough guys because they make massive migration journeys, some inconceivable for creatures so small. He also hung two other billboard-sized warbler paintings over the subway entrance at 157th Street—a difficult installation performed on a lift in the middle of the night in winter, after ripping down calcified signage for a drink called "Energy 69." From that vantage, he witnessed the Rockefeller Center Christmas tree being driven down Broadway.

When I asked why George was drawn to warblers, he said, "I feel as though warblers in particular can be important ambassadors for conversation, like a gateway drug." He's noticed it's often the spring migration of warblers, especially in the eastern United States, that gets people interested in and then hooked on bird-watching. "This curiosity leads to learning about these birds, and the first step to caring about something is knowing about it. The fact that these birds live in multiple countries also encourages a broader view of conservation. If a birder in Ohio wants to see the return of 'their' warblers, they have to also be invested in preserving habitat in Central and South America and the Caribbean."

There are three billion fewer birds in the United States and Canada than fifty years ago. That's a staggering 29 percent reduction in population. We know this because of the bird-watchers who, having caught the spark, and out of devotion, submit their observations to databases and help carry out population surveys year after year.

Conservationists say the birds are dying because of pesticides, habitat loss, urbanization, and problems caused by development, as with glass-clad high-rise buildings like the Circa, on the corner of 110th and Central Park West, in front of which, especially during migration season, dozens of bird carcasses can be found on the sidewalk each morning after window strikes. Some researchers also point to climate change. In fall 2020, scientists reported a mass die-off of up to a million migratory birds across the American Southwest and Mexico. Many had little muscle mass, seeming to have dropped dead from the sky mid-flight, like the proverbial canaries in a coal mine. The ongoing wildfires in the West were considered a factor. Of the disappearing birds, warblers are among the worst-hit groups and, therefore, the most commonly represented in the murals. Their population has declined by 617 million in the last half century.

George feels the barrier of entry to bird-watching, and thus a connection to wildlife, is low. I'm not sure that's the case, though I want it to be. In his view, one doesn't have to go to a national park or a far-flung part of the planet to see birds. They are everywhere, and one's connection and concern for them leads one to care about environmental issues everywhere. "Our birds are their birds, too. It's all connected," he says.

2021

Pray for Us, 125th St. and Frederick Douglass Blvd., Central Harlem

LESSONS IN SURVIVAL

Because of our past, which is not really past so long as white supremacy persists, my extended family chose, in the summer of 2018, to celebrate the Fourth of July at the newly opened Harriet Tubman Underground Railroad State Park and Visitor Center on Maryland's Eastern Shore. The park abuts a wildlife refuge between the place where Tubman was born and where she grew up, and so the Blackwater swampland she knew— the plantation, the canal where she floated timber, the marshes where she checked the muskrat traps of her so-called masters, the territory where she sang "Swing Low, Sweet Chariot," to signal that it was time to run—is preserved. We felt that Tubman was the best and bravest model of independence our nation had to offer, since her individual freedom meant nothing to her without the freedom of her community.

I photographed my kids, ages five and seven, with their cousins at the entrance to the Tubman Center. I was struck by three forces of nature that day. The first was the indomitable spirit of "the Moses of her people" herself—the woman who risked her life thirteen times in ten years by returning south to lead seventy-some enslaved humans out of bondage. The second was the peril of sea level rise—the marshy landscape Tubman navigated as a brave conductor on the Underground Railroad is slowly going underwater. The third was honorific; my nephew Albert—named after my brother Albert; named after my father, Albert; named after my grandfather Albert, who was killed but whose spirit lives on. Thus, at one historic site coalesced three different examples of resilience.

My grandmother Mabel Sincere Raboteau fled the coastal town of Bay St. Louis, Mississippi, and the terror of Jim Crow, along the northern pathway of the Great Migration, to Michigan, in order to save her life and the lives of her children. The youngest of them was my father, Albert Jr.

He was still in her womb when a white man shot and killed her husband, my grandfather, practically for sport. I probably don't need to tell you that Albert Sr.'s murderer went scot-free.

The details are fuzzy. The man's name was Summeral. He owned the icehouse in the Bay. A fight had gone down in the icehouse. Mabel's friend Nini was threatened. My grandfather went to Summeral's house to confront him. Summeral wasn't home but his wife was. She relayed my grandfather's anger to her husband. The next day, Summeral went to the grocery store where my grandfather worked as a clerk, and shot him to death. There were no witnesses. Summeral was arrested. He claimed self-defense. He was not prosecuted. With such ease was my grandfather dispatched down South in 1943.

The courage it took Mabel to escape from harm's way and start her life over was no less extraordinary for being such an ordinary African American story. Her fugitive ways were also the gift of improvisation; the same kind of talent that could draw low-down music out of an upside-down bucket rigged with the handle of a broom, and a single string. What other gutbucket choice did she have but this? As Mary Annaïse Heglar points out in her vital essay, "Climate Change Ain't the First Existential Threat," there's a shortsighted arrogance to the environmental movement when it claims ours is the first generation in history to face annihilation.

Faced with the horrors of climate change, how do we define or redefine the term "resilience"? What do we mean when we name a person, place, or thing "resilient"? When I hear that word, I can't help thinking of my grandmother, and of both the Great Migration and the Underground Railroad, as prior examples of organized retreat—albeit from a different hazard to life, liberty, and the pursuit of happiness than the rising sea.

Public art in the form of climate signs has made me reckon with the existential and financial threat posed to the United States coastline by sea level rise. I'd followed the signs out to Staten Island, where managed retreat has taken place, with my friend, Mik. Some folks consider managed retreat to be our best defense, given the scale of the problem. This approach calls for withdrawing rather than rebuilding after disasters, and would include government buyout programs to finance the resettlement of homeowners from vulnerable areas.

Of course, the climate crisis has worsened in the short time since Mik and I wandered the city, pointing to the challenge of a story that cannot keep pace with its subject. The picture is moving at too many frames per second. In August 2019, Indonesia announced plans to change its capital to Borneo from Jakarta, which is sinking beneath the sea. That same year, the Intergovernmental Panel on Climate Change (IPCC) released a new report showing that low-lying coastal zones, home to 680 million people—about 10 percent of the world's population—are at severe risk of increased sea level rise, extreme weather, and more frequent and stronger storms. More than half the world's megacities, and 1.9 billion people living on the world's coasts, are in grave danger, and several cities are already disappearing underwater.

New York is the center of my universe, but this is not only a New York story. It's Mik who taught me that sometimes the clearer picture is the one where the camera's pulled back. I want my eye to be big, like the eye of the woman in the Bushwick mural.

Is there a wide enough aperture to show what the climate emergency looks like now? Or is the trick to look at the past? How do we ensure that a strategy like managed retreat doesn't result in unjust displacement? There is a pernicious history in this country of the forced movement of people of color, from chattel slavery and Native American removal to Japanese American internment camps, segregation, redlining, urban renewal, "slum" clearance, and real estate exploitation. Given our track record, it wasn't surprising to learn from a report in the journal *Science Advances* that the selection of properties for federal buyouts among the homes in flood-prone areas bought and demolished by the Federal Emergency Management Agency (FEMA) in the last twenty years had as much to do with income as danger. The factor should be danger.

At the Harriet Tubman Center, marsh grasses have sensibly retreated to higher ground to survive the water's rise. Aerial photographs taken between 1938 (twenty-five years after Tubman's death) and 2006 (eleven years before the visitor center opened in her name) indicate that five thousand acres of the Blackwater refuge's marsh have become open water. The ecological system's ability to adapt to climate change demonstrates the same hardiness both my grandmother and Tubman enacted: reroute or risk death.

The engineers of the historic landmark site studied the implications of marsh subsidence before embarking on construction. The project leaders weighed the risk of climate change against that of consecrating the precise ground that Tubman survived, traversed, and repeatedly escaped. (They also expected the center would generate millions of dollars a year for the state economy.) In the end, builders hauled in tons of extra fill dirt to raise the first floor of the site two feet above base flood elevation. Despite these efforts, by the century's end, the park constructed in homage to Tubman will likely be flooded. In fact, the Union of Concerned Scientists listed the Tubman site among seventeen endangered national landmarks, long before its official grand opening in 2017. That's because the last time atmospheric carbon dioxide levels were this extreme, the world's oceans were some one hundred feet higher than they are now.

There's no longer any scientific doubt that our greenhouse-gas emissions are warming the globe and melting the ice caps, raising the level of the sea and endangering developed coastlines all over the world. The question is, how high will the water rise, and how soon? Predicting sea level rise is an inexact science. According to the IPCC, we're looking at between one and three feet of rise by the end of the century; the UN forecasts three. A recent model by the National Oceanic and Atmospheric Administration (NOAA) presents a worst-case scenario of eight. Other experts, like geologist Harold Wanless, believe these predictions are gross underestimates. The current rate of rise is doubling every seven years, he maintains—meaning, if we keep careening along this plunderous track, 205 feet of sea level rise by 2095. Although he doesn't expect we'll get that much rise, he believes we'd be wise to prepare for fifteen. The West Antarctic ice sheet is collapsing sooner and faster than predicted. There's enough glacial ice in Greenland alone to raise sea levels about twenty-five feet. Greenland lost a record 12.5 billion tons of ice in just one day in August 2019, prompting one NASA oceanographer to sound the Klaxon: "We should be retreating already from the coastline."

Although she appears in only one paragraph, late in the book, I felt affirmed to encounter Harriet Tubman in Elizabeth Rush's book *Rising*, an elegiac environmental justice–oriented meditation on sea level rise that deepened my concerns sparked by the signs. Rush deploys Tubman's story as an example of courage combined with practicality, detailing how, on her dangerous missions to lead others to freedom, Tubman marched at night, communed with God, drugged crying babies, and even held a gun to the heads of those who grew weary or wished to turn back, warning, *You'll be free or die a slave.* By this point in Rush's narrative, she's funneling toward her conclusion about vulnerable coastal communities and their adaptive strategies: true resilience means preparing for collective, egalitarian retreat. Like Heglar, whose essay emphasizes what Black people in this country have long known about building movements, courage, and survival, Rush pays respectful attention to those most at risk. She writes: "The reality is that many living on climate change's front lines are low- to working-class people and communities of color, whose relationships with the more-than-human world regularly go unaccounted for in the 'official story' of environmentalism we tell in this country."

Hurricane Katrina underlined this reality most visibly in New Orleans, where white and Black residents suffered differently. Since nearly one in three Black residents did not return after the storm, it's no longer the Black city that it was. Less visibly, the single most important factor corresponding to where toxic waste facilities are sited in the United States is the race of residents, and as a predictor of the distribution of pollution, race is more potent than income. That means it doesn't matter if we get a pay raise; those of us who choose to remain in Black and brown communities still get the shit end of the stick. EPA scientists found in 2018 that people of color in forty-six states live with more air pollution than whites.

Troubled by such inequities, Rush reveals why and where the sea is rising, who in our nation is affected, and what we might democratically do about it. She excels at redrawing our blurring edges—showing, for example, how the state of Louisiana no longer resembles a boot now that its sole is deteriorating; how big, spongy swaths of New York, Boston, Providence, New Haven, Philadelphia, Baltimore, and Washington, D.C., were developed on backfilled wetlands.

In her book *Charleston: Race, Water, and the Coming Storm*, Susan Crawford, a professor at Harvard Law School—where she teaches courses on climate adaptation and public leadership—also takes on the unequal racial burdens of rising seas, reckoning with what compounded denial, boosterism, widespread development, segregation, gentrification, white supremacy and public complacency have wrought. For centuries, Charleston has played a starring role in the nation's tortured racial history: first as a major slave port, then as a central domestic slave market, then as the

spot where the Civil War started, at which point the ratio of Black people to white in the state was around three to one. Charleston's economy
was developed on the backs of the enslaved who worked in the rice pad-
dies and picked the indigo and filled the soggy edges of the peninsula
with trash and rubble and offal so the city could grow. Which it did in
spades, most remarkably in the three decades following Hurricane Hugo
in 1989, when the local mayor harnessed national attention plus public
funding to develop the peninsula and spread outward, over marshes and
sea islands—a process that involved annexing suburbs, attracting retirees, gentrifying rampantly, and transforming a majority Black city into a
majority white one.

Discrimination persisted into the current era, during which, after the
massacre at Mother Emanuel AME church in 2015 that sent me on a
pilgrimage looking for answers in murals about how to talk to my children about being imperiled, long-simmering tensions boiled into protest
that finally brought down the Confederate flag at the state capitol. But if
you've gone through centuries of second-class citizenship and generations of flooding without being heard or helped to get to higher ground
by the city you built, as the toxic sludge pulls at your ankles, you must ask
yourself, is this progress? You must ask yourself, what do I even call the
ground on which I stand?

≡

Rising is also a treatise on language. Take the term "resilient," which we
apply to both people and the environment to describe strength. Rush
glosses the word, considering how its meaning varies for people depending on where on the shifting shore they stand. While resilience in Manhattan post-Sandy might look like the massive $20 billion hurricane
barrier proposed by the Army Corps of Engineers to protect New York
Harbor with surge gates similar to ones in the Netherlands, resilience
can suggest something altogether different in lower-income communities of color. In New Orleans's Lower Ninth Ward, or parts of Bay St.
Louis in the Gulf of Mexico, where property values were suppressed by
redlining—and where much of my father's family remained after Mabel
fled—it meant that after Hurricane Katrina, people couldn't afford to
rebuild their homes and had to migrate. To me, resilience looks like my

cousin Tracy and her husband, Charles, who, after almost drowning while praying to God in an attic with their children, Imani and Omari, perceived that the news choppers swirling overhead weren't there to save them, but rather to make pornography of their suffering—and still did not lose their faith in God. In the storm's aftermath, they picked up and moved to Atlanta, part of a mass ad hoc exodus that marked them, according to some at the time, as the "world's first climate refugees."

Resilience may also look like managed retreat; in her book, Rush investigates two coastal communities, neither of them wealthy, in New York and Louisiana. Each banded together to secure funds for relocation. After Hurricane Sandy, residents of Oakwood Beach, on the low-lying eastern shore of Staten Island, accepted pre-storm prices for their properties through the federal Hazard Mitigation Grant Program (HMGP) rather than rebuild their destroyed homes. To get those funds, homeowners had to stop thinking of retreat as defeat and decide unanimously to leave. Their houses were then demolished so that, through the process of rewilding, the land could act as a buffer against future storms. In the second case, the state of Louisiana paid to relocate and shelter islanders from the drowning bayou community of Isle de Jean Charles with HUD funds secured through a design proposal to the National Disaster Resilience Competition. Both examples demonstrate the wisdom of fleeing together, with support.

But there isn't enough money in the federal coffers to move every resident away from the risk of rising waters, and people can't apply directly for buyouts from the federal government. Instead, local elected officials must successfully navigate the Kafkaesque bureaucracy of such programs and also decide which homeowners can participate. Troubling new data show that buyouts have disproportionately benefited wealthy counties. Isle de Jean Charles and Oakwood Beach were outliers to this pattern.

As a broader solution to the thorny problem of relocation, Rush proposes we institute a nationwide property tax of one cent per square foot per year: the "Seas Are Rising and So Are We Tax." She argues that without an eco-socialist policy to address the ways in which sea level rise will exacerbate economic and social inequality (while also displacing and possibly drowning half the currently endangered species on Earth), we risk even more segregation, exclusion, and extinction. In the recent grim wake of Hurricane Dorian, for instance, 119 Bahamian evacuees were ordered off a ferry to Florida because they didn't have visas, which they did not legally need. Not only did Dorian's devastation demonstrate that the most vulnerable populations with the smallest carbon footprints are hit hardest, it also offered a nightmare picture of "climate apartheid," wherein the rich can move to safety while the poor may be excluded, refused, or killed at the borders of privileged nations like ours.

Dorceta Taylor, an environmental sociologist, explores these divisions more fully in *Toxic Communities: Environmental Racism, Industrial Pollution, and Residential Mobility*. She sees disproportionate responses to natural disasters as one example of environmental injustice. In an *Essence* interview called "Black Women Are Leading the Way in Environmental Justice," Taylor also cites Harriet Tubman as an early environmentalist with deep knowledge of and spiritual connection to the land.

> After Hurricane Harvey, Trump went to Texas and [the Federal Emergency Management Agency] gave over $100 million. After Hurricane Maria devastated Puerto Rico [a US territory] Trump basically told them to pull up their bootstraps, and it took him almost two weeks to visit. . . . He still has not actually set foot in the Virgin Islands.

Taylor's crucial inclusion of the US Virgin Islands and Puerto Rico in her definition of the American shore points to the absence of these territories as sites of serious inquiry elsewhere. The American citizens of these battered island territories surely have much to teach those of us who live in the contiguous United States about adaptive survival strategies, as do my grandmother Mabel and the Central American migrant caravans pushed in part by climate collapse to our southern border. New York City, an archipelago of about forty islands, has much in common with island nations like the Philippines, the Seychelles, the Maldives, Cape Verde, and Indonesia when it comes to sea level rise. Increasingly, because of climate change, our received ideas of strict national boundaries no longer hold. Will we, like birds, stop seeing borders? The burning of the Amazon affects all of us—affects my kids and all kids, and the kids yet to come—as do the melting of Thwaites Glacier and the decline of honeybees.

With thousands of miles of shoreline to protect along the Atlantic and Gulf Coasts, and three trillion dollars' worth of property at stake, journalist Gilbert M. Gaul argues that it's financially unsustainable in the United States to continue rebuilding along our beaches. In this context, the meaning of a term like "resilience" becomes unclear: What makes a coast resilient? How much will it cost to make it so, and who will pay?

Gaul follows the money, along with the pace of destruction, to lay bare our colossal failure at national planning. Seventeen of the most destructive hurricanes in US history have occurred this century. In 2017 alone, Harvey, Irma, and Maria resulted in more than $300 billion in combined losses. Katrina cost $160 billion; Sandy cost $72 billion. That same year, researchers for the online real estate database Zillow discovered that almost two million homes (half of them in Florida) worth roughly $900 billion could be swamped by 2100. Gaul predicts that soon we'll be walloped by a hurricane with a $250 billion price tag because we've foolishly developed so much property on barrier islands and coastal floodplains. We've done so with the encouragement of a bewildering amalgam of federal tax breaks, low-interest loans, grants, subsidies, and government insurance policies that have shifted the risk from private investors to taxpayers, distorting our understanding of that risk. Imagine this: by the end of the century, eighty million Americans could be forced to flee the coasts, redefining whom we consider refugees.

"Climate refugee" is not a legal term, though according to the Internal Displacement Monitoring Centre, seven million people worldwide were displaced by the climate crisis in the first half of 2019—*before* Dorian struck. By 2050, climate collapse is projected to push as many as 1.5 billion to leave their homes. As we're seeing, the poorest populations least responsible for the problem are the first to migrate. Increasingly, as the crisis unspools, we may have to recognize "climate refugees" as a category, acknowledging not only cross-border migrants but also those, like my cousins Tracy and Charles, who move within their own nations.

In Gaul's book *The Geography of Risk*, he shows that what we think of as our coast is a relatively modern phenomenon. In the postwar boom, many Americans indulged in the purchase of second homes in beach towns. Developers dredged and filled tens of thousands of acres of fragile wetlands, and thousands of acres of salt marsh were destroyed to make way for houses. Private agencies eventually stopped selling flood insurance because it was too expensive to cover the claims, so an arm of FEMA called the National Flood Insurance Program (NFIP) was founded in 1968 to defray the cost to the federal government. Homeowners in flood zones were required to pay for an insurance policy financing future recovery, but premiums were richly subsidized. Unlike the Dutch, who have a national tax to pay for water defenses but no government flood insurance, or Canadians, who've responded to escalating environmental disasters by limiting aid to flood victims, in America the NFIP made living in the floodplain seem cheaper and safer than it actually was. The number of residences in hazardous flood-prone areas has quadrupled since the program was founded a half century ago. Through storm after storm, Americans applied for aid and rebuilt.

Development at the coast has gone on nearly unchecked, and coastal flood claims have increased twentyfold in the last twenty years. Verging on collapse, the NFIP has lurched from one disaster to the next and is now more than twenty billion dollars in debt. Vexingly, the most generous taxpayer subsidies accrue to some of the riskiest properties. We're ponying up to pump sand in front of the summer homes of millionaires, many of whom vehemently protest the idea of "retreat."

Both Gaul and Rush cover various resilience plans and policies to address sea level rise at local levels. These include widening eroding beaches, replenishing sand dunes, restoring wetlands and marshes, raising streets and houses, designing horizontal levees, constructing floodgates, erecting seawalls, pumping out salt water, building floating cities, reforming flood insurance, competing for backlogged Army Corps funding, and more. But having spoken to enough experts, both writers suggest that these strategies are ultimately just buying time. The water is rising, and sooner or later we'll have to move inland. Masses of Americans will have to move, just as masses of people elsewhere in the world already are.

After shocking us with the sticker price to adapt the coast to the warming world, Gaul asks us a series of difficult, market-driven questions. Texans want $61 billion to protect their coast from hurricanes and floods, half of it for a barrier and levee to protect Galveston and the Houston Ship Channel. Say we pay for that—how do we also afford to protect Mobile, New Orleans, Tampa, Miami, Jacksonville, and Boston? Can we, as a nation, pay for more than seven million houses and businesses in the floodplains by continuing to rebuild? Can we pay for Miami and Norfolk and Atlantic City and also erect walls around New York City and the entire Florida peninsula? Not factoring in homeowner buyouts, fortifying stormwater systems, and other ways to mitigate flood risk, the astounding cost of building seawalls to protect US coastal cities will reach four hundred billion dollars by 2040, according to a recent study by the Center for Climate Integrity. What's the moral capital of all that money, and where do we draw the line?

"The issue isn't the Outer Banks," Stan Riggs, a coastal geologist, says on a trip with Gaul across the coastal plain of eastern North Carolina. "They can take care of themselves until the water gets them. The politics ignores all of these small, poor places on the Inner Banks." Together they explore the poor, flat, waterlogged country that Riggs dubs "NC Low" of sleepy towns with under a foot of elevation, some built and settled by freed slaves who'd worked on nearby farms, now dwindling, threatened, and ignored. Most of the focus on the rising water lay where the moneyed property was. "The towns out there on the barrier islands are fine, at least for now," he says. "But if only the rich can afford to build seawalls and widen beaches, that's not much of an adaptation. It's climate gentrification."

"Real resiliency might mean letting go of our image of the coastline, learning to leave the very places we have long considered necessary to our survival," Rush writes. Gaul focuses on the thorny issue of unsustainable cost (without mentioning the current wave of lawsuits against the fossil-fuel industry—which many believe should help foot the bill for the destruction it has helped to engender. Who is suing? Kids, farmers, fishermen, cities, and states), Rush focuses on the strategy of community-driven resettlement with government support. It is, she argues, the only approach with the appropriate humility and acknowledgment of the scale of the threat. At the end of the chapter highlighting Tubman, she writes that of all the policies being discussed, managed retreat "is the only one that calls for everyone living on the lowest-lying land along the water's edge, those who can afford to lift their homes and those who cannot, to participate." Well, yes, but only if the process is managed more humanely than resettlement has ever been managed in the past.

≣

At Tubman's home in Maryland, I remember spying ghost forests where encroaching salt water has killed off loblolly pines, a sign of habitat loss and a harbinger of flood. Is "flood" the right word? In her influential essay "The Site of Memory," about how the imagination works, Toni Morrison wrote, "You know, they straightened out the Mississippi River in places, to make room for houses and livable acreage. Occasionally the river floods these places. 'Floods' is the word they use, but in fact it is not flooding; it is remembering. Remembering where it used to be." Loblollies live long enough that some of those ghost trees may have been saplings when Tubman used the moss on their trunks to orient herself in the woods on nights too cloudy to spy the North Star.

My grandmother Mabel's values were detached from homeownership. Nevertheless, when she fled state-sanctioned white terrorism in Mississippi, it wasn't easy for her to go. Mabel loved that land, even the beaches along the Bay where she was forbidden to swim because of her race. She loved the song of the cicadas, the sea foam at the dark hem of the water, the sharp sea-salt smell in the humid air, the dusting of talcum powder below the necks of the nanans, parrains, old aunties, and cousins where they clasped her and held her tight, speaking their long-time love in Creole into her scalp, her temple, the shell of her ear; she loved the hospitality of our people. That was the land of our ancestors, who knew it by blood and by toil, though they did not own it. Bay St. Louis was her home.

In the summers she brought my father back home, down South. In his spiritual autobiography, *A Sorrowful Joy,* my father tells a story from the summer of his seventh year, around the age my sons are at the time I am writing this. To paraphrase, they stopped the car beside the beach where the hot sand burned my father's bare feet. *Y'all can't swim there, hear?* Two old white ladies rocking on their front porch across the road, sipping sweet tea. *Y'all can't swim there!* My grandmother: "My little boy just wants to wet his feet in the water." *Y'all can't swim there; you go down to Waveland; the colored beach at Waveland!* My family got back into the car and left. My grandmother: "You damn cracker bitches, I hope to God your house blows down in the next hurricane." And it did. My father saw this as "the power of a mother's curse." I never got to meet my grandmother, though we were born on the same day; she suffered dementia and died of arterial sclerosis far earlier than she should have. In my diagnosis, this country made her lose her mind.

At times, I wonder if the essence of climate denial among disbelieving Americans, most conspicuously our forty-fifth president, is an unwillingness to accept that there is anything in this world so powerful that it could overrule them. It seems to me that many Americans, especially white Americans, have been taught to believe that this nation belongs to them, all of it, and they are loath to concede that ownership. My grandmother was a refugee. She prized community over property. By cleaning the homes of white people—by dusting their bookshelves and scrubbing their toilets down on her knees—she was able to raise her three children in Michigan. They all lived well into old age. She ensured their survival by running. This required sacrifice, humility, strength, and faith. This is what Mabel knew, and she knew it from people like Harriet Tubman. When something is going to kill you, you run. Your chances of survival are stronger if there's a pathway to carry you. If there is no path, you forge one. It would behoove this nation to follow the example of people like her.

2019–2023

Cloud Dancers, 103rd St., Upper West Side, uptown C train platform

DO EPIC SHIT

One person forging a path to climate survival is Luz, whom I met at one of Angie's house parties, before the pandemic. Luz talks like this:

Do you know where to find water in an emergency if it gets shut off and you can't buy it from a store? Do you know what to pack in your bug-out bag? Tuna packets. Nuts. Mini flashlight. Compact charger. Medical supplies. Change of clothes. Water. Filter. Knife . . .

By her own admission, Luz (whose name has been changed to protect her anonymity) doesn't match the stereotype of a prepper. She isn't white, male, extremist, conspiracy-minded, militant, completely off-grid, a hoarder, a homesteader, gun-crazed, paranoid, nor a believer in apocalyptic millennialism. She does, however, live in an RV (quite comfortably, mind you), sensibly prepared for disaster. "People are surprised to see a Black woman in the lifestyle," says the fifty-year-old Afro-Latina New Yorker. "I'm a unicorn." When I met her, she'd been in the van, nicknamed Langston, going on five years.

Previously, Luz lived in a house in South Beach, on Staten Island's eastern shore. But in 2012, when Hurricane Sandy slammed the coast with a sixteen-foot storm surge that crushed homes, killed twenty-four residents, and destroyed everything Luz owned in a violent whirl of salt water that rose to the second story of the house, her faith was shaken and her life turned upside down. Her belongings went in a dumpster: books, pictures, clothes, "todo." For nearly a month, she went without power, light, or heat. So did everyone else in that flooded community close to the water. Unlike residents in nearby Oakwood Beach—who banded together to take advantage of a state program that paid them the pre-storm value of their homes to relocate, so that an uninhabited buffer zone could be made to guard against future storms—Luz had no home equity to barter for her future. The house she lived in was a rental, and the lease was in her brother's name. While the landlord knocked out the moldy walls with a sledgehammer and dutifully rebuilt them around her, Luz started burrowing down Internet wormholes, researching how to survive the next catastrophe.

The trauma of Hurricane Sandy was compounded, in Luz's case, by a recent breakup with a long-term girlfriend, a period of job insecurity, and debilitating grief from her mother's untimely death. Even now, she's too haunted by the details of that tragedy to go into it. Luz had medical insurance through her job, but it didn't cover psychotherapy. As for many Americans, paying out of pocket for mental health was a luxury she couldn't afford. The carelessness of the busted health care system only left her more enraged, vulnerable, and edgy. "I felt like a loser," she says. "I wanted to fight somebody." She found unlikely comfort in prepper videos on YouTube that supplied her with practical information about alternative energy, emergency supplies, survival backpacks called bug-out bags, and bug-out vehicles stockpiled with food and fuel. When one "bugs out," they abandon their home because of an unexpected emergency. This wormhole led her in turn to websites about RV life.

RV life wasn't just for retirees, Luz discovered. It was an organized community. Many mobile homeowners were artists keeping a low overhead while making art, and this appealed to the creative side of her, the part that wanted to detach from the grind and go on a writing retreat. There were apps that showed where to find the cheapest gas and where to camp for free. She looked longingly at Art Deco Airstream travel trailers and artfully remodeled school buses.

A liberating philosophy took shape in Luz's mind: Think big. Live small. If disaster struck again, she could drive away with her house on her back. And since she worked mostly from home, managing cases for a city agency that assessed babies with special needs, she could more or less go wherever she pleased. She didn't have kids of her own, attachments, or entanglements to anchor her to an address. With less house to pay for, she factored that she could cut back her hours at work, freeing up time to devote to writing. The more she imagined downsizing, the more empowered she felt. Maybe healing didn't have to unfold on a therapist's couch.

Resolved to remodel her way of life in the spirit of conservation, safety, sustainability, creativity, and adventure, she shared her dream to move into a recreational vehicle with her siblings. Their initial reaction was to stage an intervention. "I knew she'd been talking about getting an RV, but I didn't think she was serious!" says her brother Henry, looking back. He wondered about wild animals in the woods or truckers on meth at truck stops. Where would she park? How the hell would Luz, a woman alone, protect herself? "They thought I done lost my mind when Sandy pushed me into this lifestyle," Luz says, smiling ruefully. "Let's see how crazy I'ma look when that shit happens again."

After reassuring her concerned family members that she had security precautions in place, including financial backup should she change her mind, Luz purchased Langston, a sleek, gray, passenger-sized, pre-owned Winnebago Era the length of two cars. She tricked him out with solar panels, a cast-iron wood-burning stove, and a queen-sized bed. She ripped out a cabinet to make more room for dancing in the narrow aisle between the wet bath and the two-burner range where she would cook rice and beans. She decorated her new home with plants and Dominican dolls, stuck a Saint Christopher statue on the dash, and hung a dream catcher from the rearview mirror. Above the sink she arrayed Scrabble tiles spelling out:

THANKFUL.

When she invited her extended family for a reunion picnic at a campsite in New Jersey, they were duly impressed by her new digs. One little nephew thought the van was so cool that he wanted to move in. Nowadays, Luz spends most of her time at campsites, in rich solitude, enjoying nature.

Luz wasn't forced by natural disaster to move into a van, nor did she move across an international border. So far, she's not strayed from the tristate area. Yet it wouldn't be a stretch to call her a climate refugee. She refers to herself as such, and as an "involuntary minimalist," as well as an unlikely prepper, but she allows the label "climate migrant," too. The UN University Institute for Environment and Human Security defines a climate migrant as a person who leaves home because of environmental stressors such as drought, floods, or storms. Over the past decade, about twenty million people a year have been forced to leave their homes because of climate-fueled disaster, according to a recent report from Oxfam, mostly people from poor countries contributing the least global carbon pollution. That's equivalent to one person every two seconds.

Nobody knows for sure what the climate crisis will mean for worldwide human population distribution looking forward. "50 MILLION CLIMATE REFUGEES" was one of the messages flashing on the signs Mik and I toured around New York. Two hundred million climate refugees by 2050 is an oft-quoted, oft-debated guesstimate. A 2018 World Bank report estimated 143 million climate change–driven migrants by 2050 from the regions of sub-Saharan Africa and Southeast Asia alone. But Luz reminds me it's a mistake to think the climate emergency is relegated to the province of small island states, low-lying megadeltas, and the Sahel.

Climate stressors that have displaced people include floods killing livestock in Mozambique, drought shriveling crops in Somalia, salinity intrusion ruining rice cultivation in Bangladesh, and hurricanes destroying infrastructure in the Caribbean. Luz's story is part of the unsettling picture coming into focus and in league with other stories of climate migration within the United States that suggest the era of climate migration is already upon us. As Oliver Milman wrote recently in an unsettling *Guardian* article about Americans being the new mass climate migrants: "Millions of Americans will confront similarly hard choices as climate change conjures up brutal storms, flooding rains, receding coastlines, and punishing heat. Many are already opting to shift to less perilous areas of the same city, or to havens in other states. Whole towns from Alaska to Louisiana are looking to relocate, in their entirety, to safer ground." The profound population shift beginning to take place rivals any other in US history. It could exceed the 1930s Dust Bowl migration that saw 2.5 million people move to California from the dried-up plains, could rival the Great Migration wherein, over decades, 6 million Black Americans, including my grandmother Mabel and her children, fled to northern, western, and midwestern cities from the Jim Crow South.

While it's true that most climate migrants flee from rural areas since their livelihoods often depend on climate-sensitive sectors, like fishing, herding, and farming, it's just as true that sea level rise will increasingly affect city dwellers, like Luz and me, on densely populated coasts. The most likely cities in the United States to be submerged by century's end are Miami, Houston, New Orleans, Virginia Beach, Charleston, Atlantic City, and Boston. An analysis by researchers from Climate Central stated that this century, sea level rise could flood coastal regions that are now home to as many as 480 million people.

All of us, whether rich or poor, whether citizens of the so-called first world or so-called third, whether by hurricane, wildfire, food shortage, heat wave, air pollution, landslide, recession, drought, flood, or civil unrest, stand to be touched in some awful way by the crisis as the globe continues to warm.

What would you do if you got stuck in an attic during a flood? Luz has drilled me. *You should get a propane heater for your place,* she's advised, *plus solar panels, an inverter, and a battery pack to charge your electric devices, especially your phone. Always keep cash on you. Small bills: singles, fives . . .*

This was the kind of talk that made me clock Luz as the most interesting guest at Angie's party when I first met her, drinking shots of Del Maguey Vida mezcal. With her green army jacket, strong, thick thighs, and wide hips, she looked like she could knock out any chump dumb enough to cross her, but there was a quiet stillness about her, too, and a poet's fragility. She reminded me of a tree—strong, unshakable, rooted, receptive, and humble. As alarming as it felt to receive her advice, I knew she offered it from a gentle, loving place.

"Have you prepped bug-out bags for each member of your family? What would you do if you were stuck on a subway train and the power went out?" she asked. I admit that when Luz first questioned me like this, in a voice as intentional as her lifestyle, I felt fascinated but not grateful— maybe even a little put off—by the schooling in emergency preparedness. I didn't want to imagine the kind of catastrophe that would necessitate such planning, which is to say, I'd not yet been forced to. Then two things happened in short order that made me reconsider Luz's unsolicited advice.

First, the power went out in the subway, suspending service on seven lines. "Bobby was stuck underground on the 1 for forty-five minutes during rush hour without A/C," my friend Angela reported while our kids practiced capoeira in the Bronx. This was in the summer, during a sweltering heat wave. "Folks were bugging out. Ten more minutes and there would have been a riot." Next, Pacific Gas and Electric cut off the power to preempt wildfires in northern California. This was in the fall; the trees were dry as tinder because of lingering drought. "I'm researching generators and solar panels," my friend Danielle reported from the San Francisco Bay area. She lived in terror that if her phone lost its charge and she missed an emergency evacuation alert, she could die, along with her children, either by fire or smoke inhalation, and was beginning to perceive that her family might eventually have to move. Meanwhile, my six-year-old was drawing page after page of a comic called "Falling in Danger." In each and every panel an unnamed stick figure falls through one dangerous scenario into the next—out of a plane, off of a cliff, from the top of a skyscraper, through clouds and deep water—never landing. Suddenly Luz didn't sound so touched. Instead, she put me in mind of Cassandra, that woman of Greek mythology whose true prophecies went ignored.

Born in the Bronx to a tough Dominican-immigrant mother and a charismatic father who practiced dentistry on and off the books, Luz and her three older brothers grew up moving from one uptown apartment to the next—bouncing among six-story, mid-rise buildings in Inwood and Washington Heights and even touching down at a nice address on 158th and Riverside Drive—but never staying anywhere for long. Their parents cycled through splitting up and getting back together; money came and went on a parallel cycle. One time, they fled from their home on 200th Street because the building next door caught fire.

"Sometimes we'd leave with the stuff on our backs and set up shop fully. Some folks would call my childhood 'unstable,' but I learned the art of adapting," Luz told me. She credits those itinerant early years for laying the foundation for her current life. At eight, she watched the chaos unfold through the second-floor window of their place on Post and Dyckman, from which her mother pointed a flashlight during the '77 blackout, hollering for her sons to get inside. "People were running, pushing, yelling, smashing storefront windows in the dark, looting the shops on the strip and the bodega on the corner."

In the fourth grade, she remembers living in three different apartments. Langston is her twenty-second home. "My mother was no joke," Luz says. "She made a way out of no way. When we didn't want to move, she'd pull out the chancleta, threaten to hit us, and it was time to go. She had no English and little education, but she was a badass."

Eventually, Luz's mother left her father and remarried a Harlem super, and the family settled more permanently on 143rd and Amsterdam near the future mural of the Purple Finch. That's where Luz spent her adolescence during the 1980s crack era. "144th was crack city, where people cooked and cops made arrests. That was the hot block. The addicts looked like zombies. You didn't want to walk there," she says.

Her memory of 144th Street may sound faintly dystopic, but the picture Luz paints of Staten Island post-Sandy is patently apocalyptic: "It was straight-up Armageddon." The flooded house was her brother Henry's home. She'd just moved in with him, after the breakup with her then girlfriend. Prior to that, she had lived with her ex on 59th Street, not far from the East River, in a building that wasn't affected by the hurricane. "Had I stayed there just a month longer, my life would be very different," she marvels, without regret.

The plan was to stay at Henry's transitionally while apartment hunting for her own crib. South Beach, Staten Island, would not have been her first choice of a place to live, but her brother's house was at least familiar, a place to find her feet, hang her hat, and nurse her heart until she figured out her next step. Plus, she liked being near the water. The beach reminded her of the second home her family owned in Santo Domingo, two blocks from the pier on the seafront road where her mother, who went back and forth between the Dominican Republic and Harlem in the years before she died, *un pie aquí y otro allá*, preferred to ride out the months of winter.

As Hurricane Sandy drew nearer, the community was alerted to evacuate. Notice came late. Most of Luz's stuff was still packed in boxes down in the basement. She headed directly for her car. Henry, on the other hand, wanted to protect his domain. He predicted this hurricane would be no worse than Irene, the year before, which he'd ridden out just fine. She begged him to go with her, but he wouldn't budge. She says she would have lost her car to the surge if she hadn't driven away when she did, wracked with guilt for abandoning him. Through fierce wind, Luz headed north, for the campus of St. John's University, up on a hill where her ex-marine brother, Johnny, worked security. Meanwhile, Henry raced to the attic as the water overtook his house. Some of his neighbors drowned in their basements.

Safe on higher ground, Luz and Johnny frantically called the police in hopes their brother might be rescued, but the cops were overwhelmed. All they did was add his name to a list. As for Henry's phone, the battery was dying. He told them he'd tried to go out the front door when he realized the gravity of the situation, but the water had pushed against it so forcefully from the opposite side it couldn't be opened. Then, the water had entered in a mad rush. Now that the house was flooded, there was no exit. *If you live in a house, make sure you own a generator and keep an ax in your attic.* Before Henry's phone quit, the siblings said their tearful goodbyes, fearing the worst.

The next morning looked to Luz like the world had ended. It was eerie. Smoky. The streets were empty but for felled trees, blocking roads at every pass. She could only compare the scenery to a disaster movie or a war. But they located her brother in that ruined landscape, blessedly alive. For days, his house remained full of water. This water was dirty, possibly toxic. When they pumped it out, her possessions looked like they'd been hurled in a cosmic tantrum, wrung through a washer, dragged in the mud. Most of all, it hurt Luz to lose her books. She'd amassed a big and treasured library. Now it was pulp.

News helicopters swarmed overhead as if reporting on a war-torn country. The ATMs weren't functioning because the power was out. Credit cards were useless. In an all-cash society, those without cash were screwed. The Red Cross distributed water and food. There were twenty-hour lines at the gas station, monitored by police. People fought at the pumps for a turn to refill canisters with fuel for their generators, which could only be done on certain days. Utterly shook, Luz lost track of time. October turned into November. She had no winter coat, no winter anything. She vaguely perceived that then president Barack Obama's reelection was underway in some other part of the country but was still without power, so she couldn't watch it on TV. When it grew dark, she lit candles. She told herself, "If this shit ever comes back, I'll be the only person on the block with lights on."

Now that she knows how to hook up the solar through her inverter to run the lights if her batteries go out in the van, Luz is in a better place. Comfortably settled in the RV life, Luz talks about her special low-energy bulbs and the therapeutic benefits of living as she does. She's soothed, for example, by the sight of baby deer in the woods outside her windshield, and the sound of rain on Langston's roof. It reminds her of rainfall on the tin roofs of the Dominican Republic.

"Being in nature saved me," she confesses. "I feel very privileged to open the window blinds in the morning to green trees, wildlife, and the sound of birds. Some people associate it with homelessness, but I live like this by choice. It's not a crazy lifestyle. I consider it a blessing." When she craves a change of scenery, she can drive to a marina or the Brooklyn Heights Promenade to enjoy the view and make it her office for the day. Another benefit of the lifestyle is saving money. Luz's preferred campground in New Jersey, a hidden gem she prefers me not to name lest it become popular and thus overcrowded, costs just $132 per week. Insurance is under $200. "I feel like someone let me in on a secret, and the rest of y'all are missing out," she says, with pity.

Increasingly, she dislikes driving into the city, though once a week, for her job, she must. New York is too loud and chaotic for her now. In the summer, it's torturously hot compared to the country. She says she can't be around all that concrete anymore. And our level of consumption, as compared to hers, comes as a rude culture shock: how much water and electricity we waste, how much trash we produce, how little we reuse, repair, recycle. On occasion, as for a party or an event, she has to park on a city street for the night. She's learned how to minimize the risks on those occasions by scoping out quiet streets in the daytime, religiously obeying parking rules to avoid tickets, and sleeping with her head closer to the sidewalk side of her home in case some drunk driver crashes into Langston, as has happened twice before. Sometimes people grow suspicious of her vehicle parked in their neighborhoods and call the cops—it's mostly white people who do that, she says, "especially when they see a Black woman come out of the van." In a public lot at Midland Beach on Staten Island, officers regularly harassed her, knocking on her windows, demanding ID. Whenever she felt targeted by a police car circling Langston at a beach, park, or marina, she took to singing Chamillionaire's "Ridin'" in her head, or repeated this affirmation by Anaïs Nin as a protective charm: "Had I not created my whole world, I would certainly have died in other people's."

Eventually, she came to prefer street camping in Brooklyn, the West

Village, or predominantly Asian neighborhoods in the borough of Queens, where the people are indifferent. Occasionally, strangers are simply curious about the vehicle and want to ask her about it. Other times, their curiosity feels invasive, as when they wish to be let inside Langston to have a look around. The RV community discourages sharing too much information or showing license plates in pictures on social and other media for reasons of protection and privacy. (Some RVers with YouTube channels have been tracked and/or stalked.) A lesser violation Luz worries about is stabbed tires. No matter where she beds down, she faces Langston in a direction where she can get away quickly. Wherever possible, when sleeping in the city, she chooses to camp near a park.

In the fall, I went to visit Luz at one such spot—Crocheron Park, deep in Queens. She'd parked Langston in a cul-de-sac beneath a weeping willow, with a trailhead on one side and a fishing pond on the other. I thought, of course, of Thoreau. When I climbed on board, it was easy to see what Luz meant when she called her home a "permanent retreat." Langston was tidy as a ship. I enjoyed the mental clarity that comes from being in a spare space and the idyllic view of the fishermen at the edge of the pond, with its lily pads and slow-moving ducks.

Luz showed me how she uses water and energy—the lithium house batteries under the bed, the thirty-five-gallon tanks in the undercarriage, the shower that doubles as a closet (where she turns the water off to lather up with biodegradable soap, then on again to rinse it off), and the tiny kitchen sink. I told her I envied her light carbon footprint, her transcendental outlook, her freedom from too much stuff. Our three-room apartment was crammed with clutter, and we had a storage unit packed full with odds and ends I couldn't even name.

Luz nodded. "It's disrespectful to leave garbage," she said, referring to the principles of the conservation organization Leave No Trace. When the RV's gray-water tank fills with water she's used to wash her body and her dishes, she reuses it to water trees. When the black-water tank fills with her own human waste, she pays ten to fifteen dollars to empty it at a campground dump station. She says she's much more thoughtful about waste than she was before. "I try not to be wasteful of anything," she says. "I don't need anything extra. I just want what I need." One of her cup holders was full of acorns for good luck. By the woodstove, she'd set up a shrine: A fading picture of her mother. Oranges. Coffee. Holy water.

Perhaps it's this quality of spiritual optimism I've come to admire about Luz because it's something I find lacking in myself the more bad news I read. I find myself slow on the uptake to adapt as fast as needed, whereas Luz has adapted already. Our dependence on fossil fuel has unleashed a terrifying feedback loop, with carbon emissions heating the planet, multiplying threats, shrinking biodiversity, melting the ice caps, raising the ocean, flooding the coasts, wrecking the crops, altering seasons, increasing drought, worsening wildfires, strengthening typhoons, displacing populations. But climate change is not merely a fossil-fuels problem. It's a spiritual crisis. In this drastically changed and changing scheme, many more of us are going to have to move—not just from the comfortable seat of our physical homes and late-stage capitalism, when they can no longer shelter us, but into a new relationship with nature and with one another.

Luz is miles ahead on that path. When disaster struck, displacing her sense of self entirely, she did not give way to despair but leaned into a new life, liberated from consumer culture and domestic clutter, and found herself reborn. What if an existential threat, so long as it doesn't kill you, is an invitation to really live? For me, the revolution is a Black woman in the driver's seat of an RV named after a poet, with a sign by the electric panel that reads, "Do Epic Shit."

As for what's next, Luz's ambition is to save up enough money to go on a cross-country road trip, avoiding the red states. The West is calling her, she says. I envy her ability to fulfill her wanderlust. Not since traveling to Palestine had I been on a journey that epic. Luz wants to drive up the Pacific Coast Highway and witness the majestic redwoods. "My plan is to roam the country, see where I want to settle," she says. "Eventually, I want to own land, grow my own fruits and vegetables." She's been researching hydroponics on YouTube in the meantime, so she can practice farming in jars of water under grow lights in the van: cilantro, cucumbers, lettuce, tomatoes. She shows me where the jars will go, in a rack that will keep them from knocking around when she drives. "I'm fascinated by that prepper stuff," she tells me, her eyes wet with hope.

"I want to learn to grow from seed."

2019

Caution: Children Playing With Barbed Wire, Separation barrier, Israel-Palestine, artist unknown, photograph by Jakob Rubner

MOTHER OF ALL GOOD THINGS

For Tamar

My hosts had prepared me to be vague. I wasn't to volunteer that I'd been invited here to write about the occupation on the occasion of its upcoming fiftieth anniversary—or in other words, to do epic shit—but rather to say I was writing on life in Israel and Palestine in general. Just as Du Bois said the problem of the twentieth century was the problem of the color line, I'd heard it said, with a nod to Saïd, that the problem of the twenty-first century was the question of Palestine. The bulk of my trip would be spent in the occupied Palestinian territories, but I'd taken care to scrawl "ISRAEL" on the cover of my blank notebook above the date— June 2016—in the event my suitcase was searched.

I expected trouble getting through customs when I arrived at Ben Gurion Airport. In part this was because I'd had difficulty the first time I traveled to Israel, in the run-up to the second intifada. The problem then had to do with my middle name: Ishem. Evidently, it sounded vaguely Arabic. Like a lot of Americans of African descent, I was ignorant about my name's origin, and I was strip-searched at the airport as a result. I'd since learned that it's likely a German Jewish surname. I was ready to share this information if I were questioned again about my murky identity—that is, if I were racially profiled. I'd also popped a half a Xanax.

Like all customs agents, this poor woman looked like she suffered from hemorrhoids and would rather be anywhere else. I'm not sure what I looked like to her: An activist? A journalist? A Muslim? A threat? She examined my US passport. What was the purpose of my travel in Israel? she inquired.

"I'm here to visit an old friend," I told the agent, which was equally true. I wanted to know how Tamar, who had recently separated from her girls' father, was coping in this fucked-up realm and, if I'm honest, to process with her how I was coping in mine, which was fucked in parallel ways. Tamar had her kids much younger than I did. She was glad for that, she'd told me, for if she knew what she'd be getting them into she probably wouldn't have brought children into this mess.

Nina was now a preteenager, about the same age Tamar and I were when our friendship began. In the last year, the school Nina attended with her little sister had been targeted by vandals and arsonists, which of course was not fine at all, and difficult to reconcile with our own placid middle-school experience of harmonizing show tunes and riding bikes to West Coast Video and The Country's Best Yogurt. How had I let an entire decade go by?

"It's a long journey for just one week," the customs agent observed at Ben Gurion. "Why so short a time?" I wasn't fooled by the woman's conversational tone. I assumed the grilling was about to begin. I told her a week was as long as I could bear spending away from my two young children. Another truth—though I also felt giddy to be free, for a spell, of my husband, my children, and my country. Her face softened as she asked me their ages. "They're three and five," I answered. Just like that, she stamped my passport and let me through. "Have fun in Israel!" she called. I felt I'd been bestowed with a magic cape with two contradictory but mutually advantageous traits: invisibility and power.

I was a mother.

It was unusually hot for June, and the heat was dry at the desert's edge. The semiarid South Hebron Hills were stubbled with brown scrub and thistles and strewn with bone-colored rock. Though it was not quite summer and not yet noon, my guide, Ahmad S., estimated the temperature had climbed to 37 degrees—or, as my mind translated it from Celsius to Fahrenheit, almost 100. "Drink," the water lab technician reminded me. I lifted my canteen to my lips and, without thinking, drained it. A first-world privilege, this—to be thoughtless about water. We were at the ankles of the West Bank, far off the grid, in the cab of Ahmad's dusty truck.

Ahmad, who is twenty-nine and Palestinian, comes from a town northwest of Hebron called Halhul. He was a newlywed. His wife, Tamar told me, had been married before. Ahmad's brothers looked upon her as used goods because she was a divorcée, but Ahmad dissented from that point of view and had married her for love. With his light-brown skin, gelled hair, gold chain, slim-fitting jeans, and Nikes, he could pass for one of the Dominican guys in my neighborhood back home in New York City. Apart from Ahmad's slick look, I found hardly anything familiar in the desolate landscape; we may as well have been driving on an asteroid. The desert was bewildering to me as a city dweller, not just for its harsh quiet and vast field of vision but for its pitiless exposure to the sun. There were no buildings to offer cover or shade, no straight lines—just rolling hills of rubble and dark yellow dust. I felt jet-lagged, carsick, and ill at ease.

"Judea," the right-wing Zionists call it. The apostle Mark called it "the wilderness." I couldn't comprehend how such barren hills could sustain life. Before having kids, I'd been to Brazil's sertão, to the steppes of New Mexico, and to Andalusia in Spain, where the spaghetti Westerns were filmed. None of those deserts was as dry as this. Yet to the north of us grew the vineyards of Mount Hebron, famed for its grapes since biblical times. The foothills to the west extended into Israel. To the east dropped the Jordan Valley, where the storied river that the Israelites crossed bottoms out into the Dead Sea. In Israeli-settler parlance, and according to the Torah, God granted this land to the Jews.

We continued south, drawing closer to the area where the separation barrier peters out like the tail of an undulating snake. I'd say it was a no-man's-land, but there were people in it. From the passenger's-side window, I spied an Israeli settlement spread out on a bald hilltop like a green mirage. According to the international community, the settlement is illegal. I wondered aloud what would happen if we drove up there. Ahmad asked me if I had a death wish. A Palestinian shepherd, small as a speck, descended from another hilltop to lead his flock to a water source invisible to my squinting eyes.

"Throughout history, people always gravitate to the same places, wherever there is water," Ahmad said. "We have limited water here. This, as much as the rest of it, is the root of the conflict."

At close range, just off Road 317, lay our destination—the Palestinian shantytown of Khirbet Susiya, a ramshackle batch of tents, shacks, sheep pens, outhouses, a sad-looking swing set donated by the EU, a leaning dovecote, a solar panel array, and a stone monument to an eighteen-month-old toddler allegedly burned alive in the West Bank town of Duma the previous summer when a group of masked Israeli extremists lobbed a firebomb into his parents' dwelling. According to Israel, the village of Susiya is illegal. All of its structures are under threat of demolition by the Israeli Civil Administration, the military arm meant to oversee daily life in Palestine.

I wasn't emotionally prepared to confront the picture of the toddler's face inlaid in the monument as we rolled past it. It would have made me miss my own ragamuffins too much; it might even have caused me to cry, and I didn't want my sloppy maternal feelings to delay Ahmad's work. I turned my attention instead to one of the most prominent of the village's "illegal" structures—a big white water tank on stilts. Along with the solar panels, the tank was supplied to the people of Susiya by the nonprofit Palestinian-Israeli organization Community Energy Technology in the Middle East (Comet-ME). Comet employs Ahmad, who holds a degree in laboratory science from Al-Quds University. Its mission is to supply renewable energy and clean-water services to some of the most impoverished and marginalized people in the occupied Palestinian territories.

Including Susiya, Comet currently serves about thirty villages in the South Hebron Hills. These small hamlets are mostly composed of clans of shepherds and farmers who dwell in caves and tents, living much as their ancestors have for centuries, separating the wheat from the chaff, except that in recent history they've had the bulk of their land grabbed. More recently, thanks to Comet, they've enjoyed a taste of electricity.

I knew about Comet because Tamar works there, in development. She's the sole woman in a small team of quixotic physicist, electrician, and environmental-engineer cowboys who throw up wind turbines, water tanks, and solar panel mini-grids in the face of a military occupation that has discriminated against Palestinians for the past fifty years. In addition to reconnecting with Tamar, I wanted to better understand the imbalance of power that would make such an organization vital and to connect the long liberation struggle in the United States to the one unfolding here. So here I was, shadowing my friend's colleague in the Holy Land. Ahmad parked on the unpaved rugged road. We climbed out of the truck.

Today's task in Susiya was to test the purity of the water drawn by electric pump from a cistern of harvested rain into the tanks and then out through a network of pipes that snaked along the rocky ground, leading to taps and slow sand filters in the various tents. This system saves the village women the hours of labor it previously cost them to haul water by hand. Before Ahmad got to work, he took a long drag from his cigarette—one of the last allowed him during daylight hours before Ramadan, which was to begin the next day, or the day after that, depending on the fickleness of the moon. "In the spring, this is the most beautiful place in the world," Ahmad said. He must have read the look of misapprehension on my face: *The most beautiful in the world, this place?* "It's so calm," he said.

That seemed an odd word choice to apply to the contested territory of Susiya. The village gave off an air of impermanence, like a refugee camp or a site fabricated after a natural disaster, reminding me a little of the tent city I'd visited ten years before under an overpass in New Orleans, its population made homeless by Hurricane Katrina. In the past thirty years, the village has been displaced multiple times, making it an international symbol for pro-Palestinian activists of how Israel maintains brutal control over much of the West Bank by confiscating land. Susiya falls under the designation of Area C, as does over 60 percent of the West Bank since the 1995 Oslo II accords—disputed land overseen not by the Palestinian Authority but by Israel's military. In a larger conversation comparing Israeli expansionism to Manifest Destiny in the United States, Tamar described Area C to me as "lawless, like the Wild West."

Susiya has existed since at least 1830, but its Palestinian residents have been locked in a legal battle over land ownership since 1986. That's when archaeologists unearthed a sixth-century synagogue nearby, with Hebrew lettering on its mosaic floor. The Palestinian villagers were evicted, their land expropriated, and the site turned into an Israeli national park run by settlers. Palestinians are prohibited from entering the park even though its grounds also include the remains of a mosque and the caves that people from Susiya once called home. When Susiyans relocated too close for the comfort of the expanding Israeli settlement (confusingly named

Susiya 4 Ever, Susiya, Palestine

Susya, as if to reclaim Susiya), they were again expelled. In the early 1990s, Susiya's Palestinian villagers were herded into trucks by Israeli soldiers and deposited fifteen kilometers to the north under cover of darkness. Though some families scattered after being exiled, other stalwarts returned to their land, prompting escalating settler violence.

Susiya's residents were ousted yet again during the second intifada, in 2001, in retaliation for the murder of a Jewish settler from a nearby outpost—this time under the pretext that the village posed a security hazard. It didn't matter that the victim's killer didn't come from Susiya. Susiya's caves were packed with rock, its sheds demolished, its cisterns filled with debris, its olive orchards uprooted, its beehives smashed, its fields scorched, its livestock buried alive in bulldozed pens—to say nothing of the shepherds beaten and killed while tending their flocks. Many families fled to the nearby town of Yatta. What had once been a community of about eighty families dwindled to thirty, leaving Susiya even more vulnerable to attack. Under international pressure, Israel's High Court of Justice stopped the demolitions but never ordered the Civil Administration to accede to Susiya's reconstruction.

At the time of my visit, the state of Israel claimed that the roughly 350 Palestinians who persisted in the area of Susiya were trespassers because they'd erected their tents without the required permits. And yet when villagers have submitted master plans to rebuild, their requests are systematically denied, as happens throughout Area C—meaning, the people of Susiya live both hand to mouth and at the brink of ruin.

"This entire region is under stress," is how Ahmad put it, while grinding the butt of his cigarette under his heel. "You sleep poorly when each night is spent dreaming of the soldiers who may arrive the next day to crush your house."

How could he call this nightmare calm?

We followed a line of laundry strung from one of the stilts below the water tank to one of the poles supporting a nearby tent. The tent's canvas walls were held down with tires; its floor was poured concrete. Inside, Ahmad greeted a middle-aged woman who served us tiny glasses of bitter coffee while three barefoot children looked on shyly from behind a stack of thin mattresses swarming with flies. I handed each of them a ballpoint pen. (Tamar had suggested I pack school supplies.)

"I hope those are magic pens that can write in English," their mother said dryly in Arabic. I loved her for offering that chestnut. It suggested this place wasn't a dead end. A clutch of chicks skittered about at our feet. Outside: the cry of a rooster, the bleating of a lamb, the blowing of saffron-hued wind.

Ahmad knelt by the water filter in the kitchen area and filled a test tube to check for contaminants while chattering good-naturedly about pH balance, microbes, chlorine tablets, and the generally high quality of this water. If all was consistently well maintained, he said, the product was as pure as what you would find in the municipalities. "As pure as what they drink in the Jewish settlements?" I asked. I'd heard that the settlers' untreated wastewater sometimes flushed down from their hilltop septic tanks into the Arab villages in the valleys below, poisoning groundwater and springs.

"Just as pure," Ahmad said with pride.

I was struck by the mundane way he went about his job, as if it wasn't forbidden by Israeli authorities. I wondered at the level of risk Ahmad undertook to ensure this most basic need. I was also struck by the coziness of the tent and how flimsily it would buckle under the force of a bulldozer. I thought of the wolf's line in "The Three Little Pigs" (a story I sometimes performed for my children before bedtime): "Then I'll huff and I'll puff and I'll blow your house down!" This house was made out of Styrofoam, plywood, nothing, and brick. A sheet divided it in half. Its few possessions were impeccably ordered: a handful of tin pots and a two-burner gas stove, a broom, a nargileh pipe, an electric fan, and—most surprising of all, given the modest conditions—a box TV set blaring an al-Jazeera tribute to Muhammad Ali.

The boxer's young face filled the screen with dazzling braggadocio. He had passed away the day before. Ali seemed simultaneously out of place and precisely at home in the West Bank. June 1967 is a watershed moment you hear referenced all the time in Israel-Palestine: it's when Israel captured the West Bank, expanding its territory. It's also when a US court found Muhammad Ali guilty of draft evasion for refusing to serve in the Vietnam War, stripped him of his World Boxing Association title, and barred him from boxing. He had concluded that the US government was more his enemy than the Vietcong who "never called me nigger, they never lynched me, they didn't put no dogs on me, they didn't rob me of my nationality, rape and kill my mother and father." I don't know if the TV special on al-Jazeera went into all that history, but Ali was one of the few and first Americans of note to declare unqualified support for the Palestinian struggle against Zionist settler colonialism. That's why seeing

Muhammad Ali in that tent made me think of his struggle for the rights of full citizenship in direct relation to this woman's.

"Ask her what she'll do when they come to destroy her home," I begged Ahmad before we left. The woman adjusted her head covering and gestured through the tent opening, into the glare of daylight, at an unlikely rosebush I hadn't noticed growing in the rocky soil, bedecked with bright pink blooms. It reminded me of the mysterious flower in yet another children's story, *Le Petit Prince*. I thought of the Fox's line in that story: "It is the time you have wasted for your rose that makes your rose so important." I intuited this woman was the gardener. Ahmad translated her answer casually on our way out, as if it should have been obvious:

"She says she will stay right here and rebuild."

In fact, the word she'd used for her perseverance was "sumud." It means steadfastness, but it's also a political ideology that's developed in resistance to the occupation since the 1967 Six-Day War. An icon of sumud often portrayed in Palestinian artwork is the figure of the mother.

With the image of Ali flashing on the screen, I liked the idea of the mother as a similar source of strength.

The temperature mounted as the morning wore on. We scrambled from tent to tent, hounded by a loping, red-eyed desert dog. Ahmad recognized the animal from his last visit but suspected it had since gone rabid.

I can't say if it had rabies—it wasn't frothing at the mouth—but I can say I didn't want to get bit. Maybe the dog had gone mad by circumstance. Certainly it looked unhinged and toothsome, and so we avoided it. In each tent, the ghost of Muhammad Ali graced us on a Comet-powered television set. Nobody was watching, per se; as at my mother-in-law's house in Queens, the TV was mostly background noise. But it also conveyed the shepherds' connection to the wider world. *I'm the most recognized and loved man that ever lived, cuz there weren't no satellites when Jesus and Moses were around, so people far away in the villages didn't know about them.*

Ahmad went about checking the water meters to estimate daily use. The water crisis is rising for the entire Middle East due to increasing desertification, but here, in the poorest communities, the problem is most pronounced. Back home in the United States, another water crisis inflected by systemic racism was unfolding in Flint, Michigan, predominantly Black, and also poor. I could not help thinking of Flint, here in Susiya. Ahmad spoke in liters and cubic meters, throwing out statistics like the scientist he was. The daily allowance for domestic use by a family of five to ten people was no more than 200 liters, he explained, though the World Health Organization recommends 100 liters for just one person.

Even without a grasp of the metric system I did understand that Ahmad's figures adhered to a stark and troubling scale that measured not just water consumption but relative human worth. In the remote communities of the South Hebron Hills, the average person has recourse to as little as 20 liters of water a day. That's far less than your average Palestinian, who (according to the Palestinian Water Authority) consumes 73, which is in turn less than half of the 183 daily liters consumed by your average Israeli. In fact, the Israeli settlements of the West Bank receive almost limitless supplies of water through Mekorot, Israel's government-owned water company. *Haaretz* reported in 2012 that Israel's 450,000 settlers used as much or more water than the total Palestinian population of about 2.3 million.

"It is not only that they have more water than us but that they have stolen our water," Azzam Nawajaa insisted when we visited his tent. Azzam was a shepherd in his fifties with skin as tough as leather, a trea-

sure trove of classical Arabic poems committed to memory, a sophisticated understanding of the area's inequitable water use, and a sense of
outrage about its political leverage. He wore a red-and-white-checked
kaffiyeh. His sun-creased eyes blazed when he spoke. The walls of his tent
and the water tank outside it were festooned with Palestinian flags. Following the Six-Day War, the state of Israel banned this flag in Gaza and
the West Bank, and a later law banned political artwork composed of its
four colors: black, white, green, and red. Though the ban was lifted after
the Oslo accords, Azzam's flag still signified resistance. One of his wives
served me a glass of very sweet tea as Ahmad tested their family's water.
Suddenly, I was overwhelmed by the grandeur of her hospitality—how
many precious milliliters did the glass contain?

"Have you seen how green it is up there by their nice villas?" the
shepherd asked me, pointing out of the tent's mouth to the Jewish
neighborhood accross the wadi about a kilometer away, one of five satellite outposts comprising the settlement of Susya. Its greenery made
it easy to spot, as did the red-tiled rooftops of its homes, its cell phone
tower, and the symmetrical utility poles connecting its overhead power
lines. Azzam described its amenities. "They have lush gardens, watered
lawns, and irrigated farms." He speculated they might even have swimming pools. I agreed that it looked like an oasis. In truth, it looked to
me like a balmy patch of Southern California dropped into the set of
Mad Max.

"Have you seen our garden?" Azzam asked by way of contrast. He
pointed first at a rosemary bush inside a car tire, and then downhill at a
stand of parched olive trees growing at strange angles from the rind of
the earth. "They get water from an Israeli pipe that is prohibited for us
to use. We're forbidden to access twenty-seven of our cisterns since they
took our land, and we're forbidden to build new ones."

Azzam explained through Ahmad that in Susiya, they reuse the water
for cooking, then washing, then watering plants—not a single drop is
wasted. Most years are so dry that there's not enough rainfall to fill the
few accessible cisterns, and the community must supplement its supply
by buying water from Israel at a high price. A tanker truck that delivers
this water requires a permit and is forced to take long detours to avoid
Israeli military checkpoints and roads off-limits to Palestinians, result-

ing in further hikes to the cost of water. "Water sucks up a third of our income," Azzam lamented.

Ironically, the water Azzam must buy from Israel comes from within the West Bank. Shared water sources in the slowly depleting Western Aquifer Basin have been under Israel's complete control since the Oslo accords. The mountain aquifer is the main source of underground water in both Israel and the Palestinian territories, but how Israel distributes that water is grossly imbalanced. This is what Azzam meant when he said they were stealing the water—not just on a small scale but on a staggering one. A recent Amnesty International report revealed that Israel restricted Palestinian access to the aquifer's water while siphoning nearly all of it for itself. A recent inventory by the United Nations indicates that Israel's extraction was at 94 percent, leaving the Palestinians with a mere 6 percent.

We said goodbye to Azzam. Again, the mad dog: it stalked Ahmad and me with slow-moving thighs and a stare as blank as the sun on our short walk to the neighboring tent.

"We feel despair," a man named Nasser Nawajaa told me inside. He sat cross-legged on an unraveling floor mat and invited me to sit. "I hope we'll find some measure of justice through your pen."

Nasser is Susiya's unofficial spokesperson, an activist accustomed to talking about the local conflict to the international press. He was born some thirty years before, in one of the caves claimed by Israel as part of the archaeological park. Like Azzaam Nawajaa—the two men share a surname as members of the same extended tribe—the subject of water consumed him. Our meandering discussion of life under the occupation kept running back to it, like a river to the sea. I felt my lips grow chapped just listening to Nasser talk. He still mourned the cisterns destroyed in 2001 with fresh hurt, detailing how they were packed with bulldozed dirt, poisoned with rusty scrap metal or the corpses of animals, "raped" by excavator drills operated by the Israel Defense Forces (IDF). He spoke about the shameful lopsidedness in the basic quality of life.

"For one thousand liters of water, we pay five times what they pay in Israel. Meanwhile, their water supply company funnels a pipe straight to the settlements through our land," Nasser said. "We asked Mekorot to give us an opening in the water pipe. We told them we'd pay for it, even though it's ours. They said, 'No. You're an illegal village.' They have all the water and electricity they want, even though it's they who are illegal. Let's put aside the international community that says so. Even according to Israel, these outposts are illegal." Nasser referred to the peace treaty reached twenty years earlier in the Oslo accords, after which there were to be no new settlements built. The number of settlers had tripled since that time. "It's illegal for Mekorot and the Israel Electric Corporation to supply outposts like this, and yet they do it all the time."

I asked Nasser how access to clean water and electricity through Comet had affected his family. He gushed about how it had made life easier but emphasized the differences in Susiya (the Palestinian village) versus Susya (the Israeli settlement). "The revolution of electricity is like a river that can't be stopped. This has given our dark life more light. Our children can study later, we have electric outlets to charge our cell phones, and it's made a small revolution in the lives of the women. For example, they no longer have to carry water or shake a goat's gut full of milk for four hours to make cheese. We have electric butter churns now, and the Internet.

"But there's no way to even compare what kind of power we have here versus what they have there. They're on an electric grid. We're still begging for permission to crawl out of our caves and work our land, when it's clear they'll never give us permits to live here. Nobody wants to live in a cave. I'm asking for the right to exist in the twenty-first century, where people have already been to the moon and sent satellites to other planets."

"What do you do with your anger?" I asked Nasser.

He looked thrown from his script. "It's hard to keep it swallowed up inside. For some people, it spills out," was all he could say.

I pictured the jets of water surging from fire hydrants wrenched open by my neighbors on the streets of upper Manhattan so the kids could keep cool in the dog days of summer; my children's little hands splashing happily in the spray, the reprieve from the fiery heat, the overflowing gutters, the gallons upon gallons of water, the lamentable waste.

Outside Nasser's tent, the dog had curled up inside a rusted-out car body, where it bit at the fleas colonizing its ribcage. Ahmad scooped up a rock to hurl if it pursued us again. I must have looked concerned for the animal. "It's part of our culture to throw stones to protect ourselves when we feel afraid," said Ahmad, apologetically. His work was done for the day. As we passed the toddler's monument on our walk back to the truck, I forced myself to confront the child's face. "The children are always ours. Every single one of them. All around the globe," James Baldwin said.

Ali Dawabshah was his name. The picture on the stone caught him at that most charming stage—no longer baby, not quite boy. Fat cheeks, alert brown eyes, a peek of milk teeth behind his easy smile. I tried not to imagine his mother rocking his charred corpse, the sound of her keening, the weight of him in her arms. Heavy? Light?

"My son started talking at eighteen months," I blurted. I remembered when signifier joined signified on my son's tongue, the magic when babble became "ba" became "bus." I'd grown re-enchanted with the world when he learned the names for its many things. It slayed me that this child would never speak in sentences, that Ali's mother would never take delight in the sweet and silly things he had to say.

Ahmad didn't answer. He'd hung back to attend the movement of the dog as it slinked toward us with a growl savaging its throat. Instinctively, I stepped backward. I felt grateful when Ahmad threw the rock, and guilty when it struck the creature's hindquarter with a wretched thunk. The dog retreated to whimper behind a pile of tires. The man's shoulders sank. I pitied them both.

Ahmad admitted to feeling a little low—not because of the conflict in Susiya and throughout the West Bank, though that was cause enough for depression, but because tomorrow he'd start a month of fasting. The fatigue he knew to expect during Ramadan exhausted him already. Impertinently, I asked Ahmad how his body could take it. This was a question I would ask in many different ways of many different Arabs that week in Palestine—though by "it" I meant more than Ramadan.

Ahmad chose his words carefully. "We have resources deep within ourselves, like a hidden spring. We draw from this to keep going even when we have no fuel."

Then, because he could see I was dehydrated, he pulled out a peach and told me to eat.

≣

Later that Sunday, I went to the Jerusalem Day parade. Tamar, a longtime resident of Jerusalem and a secular Jew, had zero interest in joining me for the festivities. "Try not to get trampled by the mob," she warned. She stayed at home in her mixed Pat neighborhood to plan an upcoming iftar at Comet. There, her Palestinian and Israeli coworkers could break the Ramadan fast together with some of the folks from the villages they empowered. Her activist boyfriend, Guy B., escorted me to the parade instead, walking fast as a rabbit along King David Street toward the Old City with his video camera in hand. He was a gentle, soft-spoken man, fiercely intent on social equality. Not that Guy needed my approval, but I liked him as a partner for my principled friend.

We were going to witness the parade wherein thousands of ultra-nationalist Jewish celebrants proceed through Jerusalem, ending with a dramatic push through the Old City's Muslim Quarter. The city was divided after the 1948 Arab-Israeli war, with the west controlled by Israel and the east by Jordan. Jerusalem Day commemorates the city's "reunification" in 1967, when Israel conquered East Jerusalem, but it's as much a provocation against Arabs as a celebration of unity. Over a third of Jerusalem's population is Palestinian, but they're not invited to the party. Most of them know to stay inside or risk getting beat up or harassed. They mark the same date as the naksa, or "setback," commemorating their displacement after Israel's victory. Just as Susiya represented two parallel universes on one ground, so did June fifth represent two antipodal holidays on one day.

Guy referred to the annual tradition of "the March of Flags" as "the March of Hate." He's an activist with Ta'ayush (it means "living together" in Arabic), a group that uses nonviolent direct action to fight for Palestinian rights. Earlier this year, he was arrested in connection with his activity and spent some time in jail. As with the Black Lives Matter Movement in the United States, filming abuses of power is one of Ta'ayush's most effective tactics in battling state-sanctioned violence.

We trekked past Prime Minister Netanyahu's residence on Balfour Street, winding our way through the thickening crowd toward the nerve center of the Old City. After I'd visited shepherds in the timeless desert, the swift transition back into a fast-paced city gave me culture shock. Here, lining a pedestrian mall, were gelato shops, hair salons, art galleries, boutiques. I chased Guy through posses of people beating drums and waving blue-and-white Israeli flags. They were mostly adolescents—Zionist youth groups, yeshiva boys, settlers, settler sympathizers, and messianic types bussed in from across the country. Riled-up children cued for the coming apocalypse. They passed out stickers that said, "KAHANE WAS RIGHT," in reference to the late, infamous orthodox rabbi Meir Kahane, a member of the Knesset who endorsed annexing the West Bank and the Gaza Strip, along with the expulsion of Palestinians and a lot of other anti-Arab ideas that eventually brought Israel to outlaw his hateful political groups. In spirit, Kahane reminded me of Republican presidential candidate Donald Trump, who back home was finding favor by scapegoating immigrants and Muslims while agitating for a wall to be built at our southern border in an effort to "make America great again." (The response to that slogan among my Black family and friends: When was America ever great?)

Although Kahane fell from grace thirty years ago, his writings have carried on as foundational texts for most of today's militant and extreme-right political groups in Israel. Some of Kahane's manifestos were on sale at the parade, including *They Must Go*, a screed whose title pretty much speaks for itself. I felt disturbed by this nationalist display. If the bigotry of Meir Kahane could have influence here in Israel, among a victimized people with especial insight about human suffering and the need for tolerance, then surely Donald Trump could have influence in the United States among illiberal people with an investment in restoring the white supremacy at our country's foundation. He said those of us who lived in inner cities were "living in hell." One of the many new super PACs supporting Trump's alarming bid for the White House had just dubbed itself "Rebuilding America Now."

In what seemed to me at that moment a similar appeal for restored glory, many parade-goers wore T-shirts silkscreened with images of a rebuilt Jewish Temple, prophesied in the Book of Ezekiel as the eternal dwelling place of the God of Israel on Jerusalem's Temple Mount. The site of the Temple Mount, or Haram al-Sharif ("the Noble Sanctuary"), as it is known in Islam, was not far off. It was just through the Damascus Gate, inside the Old City, conquered by Israel on this day in 1967. There

lies the holiest site in Judaism, the Foundation Stone, where Jewish tradition holds that Abraham prepared to sacrifice Isaac, and where Jacob
laid his head and dreamed—the pillow of stone where heaven joins the
earth. This rock is just as sacred to Muslims for being the place where the
Prophet Muḥammad is said to have ascended into heaven, enshrined by
the Dome of the Rock that sits on the ruins of the first two temples, like
Susya on Susiya.

I shielded my eyes and looked toward that golden dome, the most
eye-catching thing in Jerusalem's cityscape, made dazzling in the evening
light. The atmosphere grew more charged as we approached that mighty
power source. The March of Flags would start at the crenellated gate and
culminate at the last remnant of the Second Temple—the Western Wall.
The masses sang and chanted so loudly their voices grew hoarse. *The
eternal people do not fear a long journey*, they belted. *Jerusalem of Gold.*

I remarked that it felt like a pep rally, one where everyone really
believed in the team. "They're brainwashed," Guy said with great pity.
Then he directed his lens at a unit of armed Israeli soldiers who'd blocked
off a street entrance to an Arab neighborhood with a tank that would use
its cannon to spray their bodies and homes with "skunk water" if they
dared try exiting to protest the march. Skunk water is a putrid-smelling
form of nonlethal crowd control developed in Israel to keep demonstra-
tions in check. (Several police departments in US cities, including St.
Louis, are reported to have recently bought it in the wake of uprisings
and protests against police brutality.)

"That's the smell of the occupation," said Guy. "It's worse than a skunk.
It stinks like raw sewage and rotting corpses. It doesn't go away for days.
It can make you sick. The term Israel uses for this pollution is 'sanitize.'"

I thought of the fire hoses police turned on nonviolent protestors,
including children, in Birmingham in 1963. The pressure from their jets
was strong enough to peel bark from a tree. "Why don't they just use
water?" I asked Guy.

"Not cruel enough."

One of the police noticed Guy filming and placed his hand on the
grip of his assault rifle. We moved on, weaving into that amped young
crowd. "Get ready. It's about to turn ugly," said Guy. He braced himself
to record the slogans, jeers, and acts of vandalism he'd witnessed at past

parades: *Muḥammad is dead; the third temple will be built and the [al-Aqsa] mosque will be burned; death to Arabs*, and so on.

"Imagine they are storming through your neighborhood, and you aren't allowed on the street, or to run your shop, or leave your house, and they are chanting death to you and your prophet," Guy said. I acknowledged it would be hard for me, in that scenario, to turn the other cheek. Someone shouldered him, hard—a teenage girl.

"Why don't you point your camera at the Arabs throwing stones?" she spat at Guy, though the only Arabs in sight were shop owners being roughly steered out of the Muslim Quarter's market by Israeli police, to clear the way for the March of Flags. I'd wanted to buy souvenirs for my kids in the labyrinthine souk, with its trinkets, sandals, ouds, doumbek drums, Turkish delights, carpets, spices, and backgammon boards inlaid with mother of pearl, but today was not to be the day. Besides, I was here to be a witness, not a tourist. Guy thought it unsafe for me to be at his side and directed me toward what looked like a safe perch from which to watch. Suddenly, he was gone, swallowed up in the noisy throng. I worried for him. Now I knew why Tamar had warned me not to get trampled. There were thirty thousand people and two thousand police officers in attendance at this evening's parade. I climbed onto the post to the side of the steps leading down into the swarming plaza of the Damascus Gate.

The revelers rode on each other's shoulders. Packs of boys danced in horas—the next generation. They seemed like their voices had just changed, and yet in no time at all, they'd be required to serve in the army. The air was suffused with testosterone and great potential, enough voltage to power a city. "What are they singing?" I asked a sympathetic Hebrew speaker below my perch. "Worship God with happiness," he translated. The air was electric. Their chorus reached a fever pitch. "God will defend us," they sang, pumping their fists. I was touched by their expressions of faith and terrified of their zeal. "We will win." It sounded to me like a battle cry.

As a mother, I wanted to shake those boys by the shoulders for being uncouth. I wanted to flip a doctrinal switch in the current of their energy, to force it back to the basic precept Jesus quoted from the Torah (Leviticus 19:34) when he said we should love our neighbors as ourselves. This

parade felt to me like the opposite of that mitzvah. As believers, as bul-
lies, the boys surged into the mouth of the Damascus Gate. It felt like a
desecration—not merely because they believed Jerusalem belonged to
them alone but because their misguided ardor was so close a surrogate
for actual joy. I thought of little Ali Dawabshah, and of my own children,
whom I tell the most important thing in life is kindness, to use their
words and not their fists; who are Black in the United States of America
and therefore also endangered. An armed policeman stared me down
with bloodshot eyes. I realized then that I was crying. What did I look
like to him? I turned my face from his because I felt afraid.

Feminist Lama Hourani was also afraid. I met her the next day, Monday,
at a trendy café in the bustling central West Bank city of Ramallah, seat
of the Palestinian Authority. She confided that her biggest fear for her
thirteen-year-old son used to be that he'd become a suicide bomber. "My
fear has changed now that he's getting older," she confessed. "These days,
I'm afraid he'll be accused by an Israeli soldier of a stabbing, or whatever
lies they will invent to justify killing him."

Compared to the rural Palestinian mother I'd encountered down in
Susiya, Lama was cosmopolitan, but like that woman, she was also tough.
An outspoken woman of fifty with a master's degree in foreign trade,
Lama wore army-green silk pants, a silver necklace, and her hair cut in a
stylish bob. She wasn't wearing a headscarf or fasting for Ramadan. Lama
claimed to be an outlier in her community for having birthed only one
child, late in life. She'd moved to Ramallah from Gaza in 2007, after Hamas
took over, and worked with NGOs to fight for Palestinian women's liberation, taking on taboo topics in a mock parliament that argued for civil—
rather than Sharia—law, debating marriageable age, polygamy, divorce,
custody and inheritance rights, and the freedom to choose one's own
partner. She joked that she was the only woman in the world proud to
be fifty. This was because at her age, seemingly no longer fitting the pro-
file of a menace to the state of Israel, she was allowed to travel through
the checkpoint into Jerusalem. But she was afraid to go, and besides, she
would have to go alone since she wasn't permitted to enter Jerusalem with
her husband, nor her child.

"I raised my son with sumud, to stay put," Lama said. You could hear the Gauloises Blondes she chain-smoked deepening her voice, but not quite masking its despair. There was a mysterious keloid scar on her forearm and another on her right hand—the hand that held the cigarette. "I'll never leave Palestine. I'm committed to the struggle, but I wouldn't blame my son if he wanted to immigrate to America or some other country. We're not free here, in our own land."

Lama continued to speak about the restrictions on her freedom of movement. Even within Palestinian borders, she feared traveling beyond Ramallah, especially at night. She had a strategy of driving at the start of Shabbat, when she'd be less likely to encounter hostile Jewish settlers out in public. She felt terrified on the road to visit in-laws in Nablus with her son in the back seat. If he made the wrong face, or said the wrong thing at a checkpoint, it might provoke a soldier to violence, and if her child were attacked, apprehended, or murdered, there would be no justice.

I felt chills: Lama was describing a version of the fear felt by Blacks in the United States, a version of my own anxiety. This was the summer of mounting civil unrest over police murders of unarmed Black men and boys, increasingly filmed with cell phone cameras and shared through social media as evidence of iniquity. As if reading my thoughts about home, Lama remarked, "Of course, it's just as corrupt in the West. In the States, you've got a prestigious constitution meant to protect everyone, but it's a lie. Everywhere you step, you see segregation. Isn't that right?"

What would be the point in pretending otherwise? In three months, my son would begin kindergarten at a public school in one of the most shamefully segregated school systems in the country. "That's exactly right," I said, before asking Lama for a cigarette.

"What? Is somebody from the West still smoking?" she ribbed.

I admitted that I wasn't a smoker, but as the mother of African American children, I wanted a smoke to take the edge off the panic. By the time my boys were teenagers like Lama's, I feared the authorities would no longer look upon them as children but as thugs. What if they wore a hoodie, or held a toy gun, or said the wrong thing, or made the wrong face or the wrong movement in the wrong place at the wrong moment? History had shown, as current events were still showing, that their lives could be taken with impunity and disregard. I thought of the white man in Mississippi who murdered my grandfather this way, back in 1943, without ever going to trial, before my dad was even born. How could we still be trapped in this ordeal? How to refuse this inheritance? Even under a Black president, time was caught in a loop where Black lives were treated with contempt.

Lama and I shared a sisterly look. Her amber eyes were heavy-lidded and strikingly large. In them, I saw my reflection: a mother of children at risk. I was prepared for motherhood from a very young age. At ten I was given the Talk, and taught how to protect my heart; it had primed me to protect my children's hearts. I know exactly what she saw when she looked at me. And when I looked at her, I saw her as a child who'd been given a version of the Talk, too. She was strong the way a survivor is strong, a woman at the edge. We regarded each other's pain in silence. After a tiny eternity, she broke that open circuit with this wisecrack: "We should both really be smoking something much stronger than these."

I couldn't help laughing. How else could we take it, except to laugh? How could Lama's body take it? I asked because I really wanted to know. How should a body best arm itself in a state of terror? It's possible I was asking Lama for advice, but other than the kindness she'd just shown, she had none to offer. Instead, she beseeched me for help with a reminder about why I'd come. As a US citizen, I had more power than she.

"You are here to take this desperation to your decision-makers. Does the US really care about Jewish self-determination? No. They wanted an ally for resources. Their main interest is energy. The main energy is in the Gulf. All of us are suffering because of that. It's not a conspiracy. You can see it." Lama's voice was strident, accusatory. What she said reminded me of Ahmad's speech in the desert the day before. Water was the source of strife, he'd said. It all boiled down to this: the fight for precious, dwindling resources. War.

I struggled to copy down Lama's tirade against capitalist greed. Her argument came hot and quick. "We can't have electricity or water without dependency on the West. We can't vote. Why? We have no sovereignty. If we were left alone to develop normally, like other countries, things would look different, but we were not. We aren't savages. I'm fed up with seeing our kids killed every day, and nobody's reacting. Why should I keep telling the story of my suffering if it's never going to change?"

I took this to be a rhetorical question. Still, I felt ashamed that my only response was to take dictation in my notebook.

"I don't trust the international community." Lama continued. "They all say settlements are illegal but don't hold Israel accountable. They're ethnic cleansing Jerusalem. How can two million people be terrorists? How is a baby a terrorist?" (Her voice broke on the word "baby.") "A stupid settler who can kill or burn kids is not a terrorist but 'crazy,' while we are called terrorists."

Lama shook with anger now and had begun to cry. I sensed crying wasn't something she often indulged in, that this breakdown was a final resort. Impatient with her emotion, or the world's indifference to it, she raked the snot from her nose and took a furious drag on her cigarette. The untouched espresso on the table before her had gone cold. The ashtray beside it was full. The last thing Lama said to me was this: "Believe me, my friend, if I knew the solution I would not be crying."

Since there was nothing I could say to defend my country's goodwill, though I could dip my pen in poison and blood, for now I put down my pen and touched the woman's shoulder. It felt like a hive of bees. Her body thrumming, electric with grief.

On Wednesday, I went with Tamar to the Comet headquarters for the iftar she'd helped to plan. It was just a dozen coworkers sitting down to supper at sundown in the desert, at a table covered by a cheap plastic cloth—Arab and Israeli guys alike, who share the belief that power is a basic human right. Compared with the despair of Susiya, the furor of Jerusalem Day, and Lama's hopelessness in Ramallah, I found this to be the truest expression of accord I'd seen in my whole time here.

The festive mood swelled in the countdown to nightfall. The cool air was perfumed by the lavender shrubs growing beyond the veranda where the table was set. Our shadows grew elongated, like figures in an El Greco painting, and then they were gone. Twilight, the magic hour. "Is it time yet? Can we eat now?" asked twelve-year-old Yusef. He sat at the table next to his father, Ali A., a shepherd from one of the off-grid communities Comet services, a place called Tuba. Ahmad S. consulted his watch, and then the slip of the moon in the indigo sky. He clapped his hands. He looked so much more energized than he had on our depressing day in Susiya. His eyes twinkled like the stars that were starting to show. It was time.

Ahmad uncovered the dishes of mouthwatering Arabic salad with tahini, grape leaves, soup, kibbe, and roasted chicken. Like a maître d', he brought out a platter of lamb and served it with flair. The other project managers and technicians were in equally high spirits as they broke their fast to dig into the feast, to make toasts, and to joke with each other in Arabic and Hebrew. Habibi, they called one another. Is there another word on earth more tender than this? "My darling." Even as an outsider, I felt inside their circle, just as I often felt as a girl when invited to Shabbat dinner at Tamar's house, around the corner from mine. Yusef felt at ease enough to joke that by my age, I should have five sons at home, not just the two. Through Tamar, I ribbed the boy back. "Why? So I could have three more like you to smack in the head for telling me how to live?" I tugged that rascal's earlobe, and he grinned.

Comet's Israeli cofounders, the physicists Elad O. and Noam D., were smiling, too. There was cause for their organization to celebrate, though this wasn't the express reason for tonight's iftar. In the eight years since Comet first electrified Susiya, it had largely succeeded in its mission to electrify the South Hebron Hills. That is, most of the so-called cave dwellers in this part of the West Bank were now connected to an alternative energy source—the wind or the sun. The team wasn't just tilting at windmills but fighting a real giant, in the guise of the occupation. Since 2008, Comet had erected 10 small wind turbines, 569 solar panels, and more than 100 household water systems like the ones I'd gone to check with Ahmad, serving approximately 2,500 Palestinians in its effort to help them remain on their land. While continuing to maintain these existing systems, Comet's next stage would be to expand its reach beyond the South Hebron Hills and establish new ones. This expansion offers an alternative to the state of Israel's. It strives not to take but to give, not to extinguish but to illuminate.

For all the prosaic nuts and bolts about generators, hard stops, and photovoltaic arrays, I'd overheard talk during a planning session about putting up a mini-grid in another community. It was hard not to consider this conversation's biblical overtones. We were in the Holy Land, after all, and these guys were deciding how much power to bestow. That level of responsibility seemed nearly supernatural. The underlying question was, how should power be put to use? Maybe because I felt a little lost in all the kilowatt talk, or because of the dreaminess of the landscape, my mind wandered to Genesis: *And God said,* Let there be light, *and there was light.*

The engineers seated at the table weren't gods, of course, but men. Nearby stood a ghostly-looking donkey, fast asleep. It was quiet now, and blessedly still, without the sound of a muezzin, or TV, or traffic, or gunfire, or anything but the wind acting on the elegant blades of the tall white Comet turbine at our backs.

Noam rolled a cigarette. Above us gleamed the waxing crescent moon. Before us on the distant horizon, in the hamlet of Shaab al-Bottum, a string of Ramadan lights blinked in the growing dark. They sparked a warm feeling in me. I felt grateful to be reunited with my old friend in this faraway place, liberated from the responsibilities and anxieties of parenthood in uncertain times. The Ramadan lights sparked a fuzzy feeling in Noam, too. He spoke to me and Tamar about the first time he saw Susiya lit up in the night like that, almost ten years before, the fatherly joy he'd felt at having helped supply the region's first light.

"Susiya is completely different today than it was ten years ago. At that time, they were very vulnerable. Activists would stay the night to protect them from getting beat up or evicted. They were at the bottom of the barrel, looked upon as uneducated. Now, they're building their position in life. They've learned how to sell their story. Usually, after you install energy, the first thing they'll do is buy a TV. It doesn't matter that I'm not in favor of that choice, or that I dislike when they squabble over who'll get hooked up first, who gets the first fridge, or the eventual air conditioner. It's their choice how to develop themselves. This is our belief."

Tamar added that Comet didn't get into issues of gender inequality. As a woman, this was sometimes hard for her—she smiled awkwardly at me, for example, when Moatasem, a technician, wouldn't shake my hand on meeting me—but she also knew it wasn't her job to make him accept it. "Empowerment means letting people live their lives as they choose to live them," Noam repeated, "even when giving people tools to argue means they may argue with you." I understood we were no longer just talking about electricity, infrastructure, or social justice. We were talking about free will and its brightest corollary—hope.

In a few days, I would return to New York, whereupon my five-year-old would tell me I was a bad mother for leaving my children for so long. The memory of this night would take the sting out of his complaint. (As would my own mother's rejoinder: "There's no such thing as a good mother.") How could I teach my kids to find this feeling if I didn't locate it myself?

Elad had somehow ripped his trousers on a coaxial cable. "The occupation did it," he declared, and everybody burst out laughing. The Muslims returned from praying, loose-limbed, to the table. It was time for dessert. We stuffed ourselves silly with qatayef, coffee, and dates until at last we were full. I gazed up at the sky, now punched through with a thousand stars and streaked with meteors. Its perfect clarity made me gasp. That night felt free in part because no veil of light pollution obscured it. We were one hour from Bethlehem, at the edge of the world, in the Milky Way. We were that far off the grid.

≡

The next morning I was rudely awakened from that state of peace. Unbeknownst to us as we ate dessert in the desert, two Palestinian cousins from Yatta, just up Road 356 from Comet headquarters, had crossed into Israel and shot up a café in Tel Aviv, killing four Jews. "Oh no," said Tamar, scrolling through her newsfeed over breakfast. Her face looked ashen.

"What happened, Ima?" asked her younger daughter, Magali, immediately on alert. A loaf of fresh bread Guy had baked sat untouched on the table.

Tamar put away her phone, unsure for the moment of what to say. Her two daughters, seven and eleven, attended a progressive bilingual Arab and Jewish school called Hand in Hand, which had been defaced and set afire by Israeli hoodlums who disapproved of its pacifist mission. (Hebrew graffiti on the walls included the shibboleth "You can't coexist with a cancer.") Tamar had to be careful about how she introduced more trauma into her girls' lives. Everyone here was suffering from trauma—Arabs and Jews alike—victims and victimizers. It was Tamar's sincere wish, and her lifework, that her kids grow up to love their neighbors. I cherished her for that mission. I hated that there were so many barriers in her path.

As Tamar readied Magali and her big sister, Nina, for the school day, I took a peek at the news on her phone. One of the victims was our age exactly, thirty-nine, and the mother of four. Her name was Ilana Naveh. I couldn't help imagining this woman's forsaken children. Who would sing to them at bedtime, and braid their hair, and kiss their bruises? As for the killers from Yatta who'd robbed them of her touch, I supposed it was like Nasser said: some bodies can only swallow so much rage before it comes spilling out.

Then, more bad news: After the girls left, Guy shared that someone he worked closely with at Ta'ayush to promote the Palestinian cause had been arrested that morning on trumped-up charges, not for the first time. Guy appeared terrified. If he left Tamar's apartment, he feared meeting the same fate. She reassured him gently in Hebrew—if they came for him, she would hide him under Magali's little bed. It was a feeble hiding place, and she was only halfway kidding. What world was this? I didn't want to believe it was so ugly, that good people should feel such routine terror, and so I went to work with Tamar's boss, Elad, with the ambition to rediscover the hope I'd felt the night before.

Elad acted nonplussed by the news of the attack in Tel Aviv, either because he's inured to the violence that incites such crimes or because of his scientific disposition. To Elad, that Thursday was like any other day. He had a job to do: standard maintenance checks of Comet-supplied power systems in the South Hebron Hills. As with Ahmad, I was allowed to tag along.

Once again, I felt carsick on our drive from Jerusalem into Palestine and down the West Bank's rangy spine. But this time, I also felt unsafe, as if I was entering a war zone. Israel's immediate retaliation for the attack in Tel Aviv included the deployment of two more battalions to the West Bank, numbering hundreds of troops; freezing 83,000 permits granted for Palestinians in the West Bank to visit family in Israel during Ramadan; and rescinding work permits for 204 of the attackers' relatives. I looked out the window at the black smudges on the landscape where settlers had set fire to the crops of Palestinian farmers, not necessarily as payback for the previous night's act of terrorism but just as a matter of terrorist course.

I asked Elad a serious question: If Comet's work installing solar power grids and clean water systems for Palestinians in Area C was illegal in Israel's eyes, then what was the level of risk? It was a personal question masquerading as a journalistic one. I wondered: In this fucked-up realm, how endangered was my dear friend Tamar?

"Legal shmegal. You're asking the wrong question," the physicist answered, brusquely. Elad reminded me of Harrison Ford as Han Solo in *Star Wars*, in part because his son's Chewbacca action figure sat on the dashboard and in part because he enacted his mission with the supreme confidence of a swashbuckler. "We're not in a land of logic," he said. "The one thing that's illegal here is the law itself."

Case in point was the highway we drove upon, Road 60. Though Elad and I could take it freely in his car with Israeli plates, the road was riddled with dozens of military checkpoints that slowed travel for Palestinians, who might be turned back or rerouted by Israeli soldiers onto dirt byroads, restricting or choking off their movement altogether. Other roads in the West Bank were cut with trenches, obstructed by earth mounds or concrete blocks. All of this fragmentation was in the name of security—to defend the Israeli settlements that are themselves illegal, according to international law.

"Try to look local," he advised as we approached one of these checkpoints at Gush Etzion Junction. Local? We were in Palestine. What did he even mean? To look more Israeli? "Less like a journalist," he clarified, "more like a mother." He told me to put away my notebook and pen, smile, and wave nicely at the soldiers. I did, and we sailed past without incident. It would not have happened so smoothly had we been Palestinian.

More soldiers were stationed with armored personnel carriers at a road that led to Yatta—"population eighty thousand; infrastructure, zero," was how Elad described the hometown of last night's gunmen. I spied a front-loader tractor dumping a hill of dirt at the crossroad as collective punishment for the murders, so that no one else from there could leave. Thousands would suffer for the actions of two. Not just this road but Yatta's every entrance and exit was blockaded. A curfew had also been imposed on its citizens. By the following week, Mekorot would cut off the already spotty water supply to Yatta and the rest of the West Bank—this in a blistering heat wave, during the holy month of Ramadan. Innocent people would die. People were already dying.

"I advise you not to stare," Elad said casually as we drove past the blockade. I found his tone disconcerting at first. He seemed nearly dismissive of the human drama of life under siege and irritated by my concern, lest it get in the way of his work. Unfazed by a sign that cautioned, in Hebrew, "This road leads to a Palestinian village. Dangerous for Israeli citizens," Elad took a sharp turn onto the steep, bumpy dirt road.

The danger I feared by now was not from the Palestinian villagers trying to eke out an existence but from the Israelis who didn't want them empowered—not by Comet, and not by me writing about Comet. We passed a man riding a bleary-eyed donkey. I thought how much my kids would love that animal, unburdened by thoughts of unexploded shells left on the ground by the Israeli army's endless exercises in the region. Suddenly, all I wanted was to go home to my kids, but the road was too narrow to turn around upon, and I respected Elad's operation too much to steer him off course. The road led us to a remote hilltop community that Comet had finished hooking up to a solar mini-grid only two weeks before. "The story should not be one of suffering but one of concrete deeds and actions accumulated over time," Elad said tersely, getting out of the car.

Elad checked the charge controller and the battery system in a shed with a ganglion of cables that fed into the villagers' concrete homes. Almost as an aside, he pointed out a nearby illegal outpost called Lucifer's Farm, with a security tower and a buffer zone; a pickup truck he suspected of smuggling Palestinians over the border into Israel to perform cheap labor; and another settlement whose residents were infamous for using slingshots to attack Palestinian children on their walk to school. He pointed out the solar panels installed by Comet, which the settlers had not yet broken and the Civil Administration had not yet dismantled, explaining how their mount angle could be changed seasonally to best exploit the sunlight. Through his careful performance of these routine tasks, I could see Elad was at heart a romantic. His actions reminded me of hexagram 9 of the *I Ching*. The taming power of the small. It signifies a time when the light is temporarily enveloped in darkness. Our responsibility in such times is to pay careful attention, accept restraints quietly, be flexible, assemble the scattered bits, and remain content with taking small steps to accomplish the great. The small good deed matters.

Next stop with Elad was a place called Abu-Qbeita, named after the extended family of thirty to forty people living there. The compound of the Abu-Qbeita family fell in an interstitial no-man's-land called the seam zone, meaning between the separation barrier and the Green Line. "Don't look for the Green Line," Elad said of the Armistice border set after the 1948 Arab-Israeli War. "You won't see it." We were in Palestine, but for all practical purposes, the separation barrier that weaves like a drunken squiggle along the map, chomping willy-nilly over the Green Line, dictated that we were not.

We sailed through the checkpoint the Abu-Qbeitas must stop at to get in and out of their home. The Israeli settlement of Metzadot Yehuda has encroached right up to their property line, backed up against the fence. Last January, settlers threw rocks that broke some of the solar panels. The culprits were kids, Mahmoud Abu-Qbeita told me, while troubling a string of prayer beads in his hands outside the electricity shed. Elad was checking the copper coils of the transformer in the shed for heat loss. He reported that the damaged solar panels weren't producing as much energy as he'd hoped. While he continued diligently to check the system for flaws, Mahmoud and I talked on his terrace.

I felt saddened by the poverty of our surroundings: an oil drum filled with trash, an upended rusty grocery cart, plastic jugs, a deflated soccer ball, a blue wash bucket, Astroturf, Ramadan lights strung in a thirsty-looking fig tree. Squint, and we might have been in an Appalachian trailer park. By contrast, the settlement on the other side of the fence looked like a well-serviced Floridian retirement community. Mahmoud and I sat on the back seat of a car being used as lawn furniture and contemplated the rolling, sunbaked hills.

Like Azzam, whom I'd talked to in Susiya, Mahmoud was a shepherd in his late fifties. He, too, complained about the lack of access to his grazing area and to water. But for him, the occupation's biggest aggravation was the checkpoint—the daily indignity of having to show a permit to go anywhere, or to come back again. The Kafkaesque procedure wore on him. It could take fifteen minutes or three hours to pass, he said, depending on the whims of the soldiers on duty. There were more than a hundred different categories of permit: permits to go to the doctor, to the mosque, to study, to visit family. Different permits for women, for men, for the elderly, for the youth. A separate permit for your tractor, for your goats. They could take your permit, if they wanted to, without explanation, because bureaucratic evil is random.

"This racist treatment angers me most for my children. I worry how it will affect them," said Mahmoud soberly. I thought I knew what he meant. I worried about how structural racism would screw with my own kids' heads; how, in all likelihood, it had screwed with them already.

Mahmoud was father to seven sons and five girls. The youngest boy was three, and the oldest had a child of his own. The children must go through the checkpoint every school day, their backpacks searched for weapons by soldiers. Those under twelve have no memory of moving through the world without being patted down. He has taught them how to be polite, to comply with "the law" so they don't get hurt, just as my husband and I will soon enough teach ours. Mahmoud's children have noticed that the settlers pass freely, without ever being checked. He indicated the place where he had wished to build a house for his twenty-three-year-old son, Bilal, when the young man recently married. But he was ordered to stop building, or the structure would be demolished. He hadn't even bothered to apply for a permit. It would only be denied, or lost in the Civil Administration's confounding bureaucratic pipeline.

"Did Bilal decide to leave after that?" I asked.

"*No.*" Mahmoud was adamant. Unlike Lama, who'd said she wouldn't blame her son for leaving, he would not condone his son's departure. "Others have left for Yatta. But not my family. We're staying right here."

I admit I found the man's obstinacy distasteful. Why would anyone willfully choose amputated possibility, jeopardy, squalor, the short end of the stick? By this point, I'd made the requisite occupation trauma tour stops at the Qalandia checkpoint, where Palestinian men crossing the border into West Jerusalem to perform cheap labor are forced to wait on line for hours to show their permits, herded through cattle chute–like cages topped with concertina wire; the casbah in Hebron, where market sellers attempt to protect their wares with a canopy of mesh from the spoiled eggs, piss, bleach, and trash hurled from the settler buildings above; the home of a man in Nabi Saleh who kept on his coffee table the tear gas canister that killed his brother in-law at a protest by exploding in his face. If I'm honest, what I really wanted to ask Mahmoud as we sat on his dismal veranda was this: *Why the hell would you raise your children here?*

Though it was the same question I sometimes asked myself, I stopped short of putting it to him. I recognized the privilege in my judgment. I couldn't presuppose he had the freedom or luxury to find a better place to go. Instead, I asked him warily if he was talking about sumud.

Mahmoud confirmed it was sumud that made him stay, but not only that. It was also that he loved his land. "Look," said the patriarch, sweeping his arm grandly at the view. I still couldn't see it, the beauty of the South Hebron Hills that he and Ahmad referred to. Maybe I was too distracted by the threat of the settlement just over the fence line, or troubled by the attacks, or drained by all I'd witnessed that week, or disappointed in him for wearing a gray hat instead of a white one. Maybe I was dehydrated, or lost, or helpless, or homesick, but all I felt in that moment was bleak.

"Do you actually believe the occupation will end in your children's lifetime?" I asked.

"I doubt it," shrugged Mahmoud. "But we're staying."

Three of his sons had come out to play. What a consolation that, in every far-flung corner of this fallen world, children discover ways to play and, in so doing, lift it up again. The boys climbed onto an old red motorcycle parked in the yard next to the electricity shed where Elad continued industriously checking the power. The vehicle was pointed in the direction of the chain-link fence. Its dusty seat was ripped. Its kick-stand sank into the soil so that it leaned heavily to the side. The oldest child sat listlessly in the seat, staring straight ahead at the settlement. The middle child pulled the throttle, pretending to drive while making a motor sound with his lips. The littlest child flipped the useless engine switch. In his hand he held a stick as an imaginary tool. I saw what he was doing with it and smiled in spite of myself. Admirably, heartbreakingly, he meant to fix the broken thing.

And who was I to think he couldn't?

≣

On Friday evening, Eid Suleiman al-Hathalin walked us past the goat pen to show us his tools. Eid was a guileless, philosophical dreamer—the older brother of Tamar's colleague Moatasem—a vegetarian, a bird-watcher, and a sculptor. He wore a periwinkle shirt and a genuinely infectious smile. The jab saws, pliers, wire cutters, and awls with which he made his art were lined up neatly on a table outside his cinder block workshop.

It was my last night in Palestine, and I was in a hilltop Bedouin village with Guy, Tamar, and her two daughters for one last iftar in the South Hebron Hills. The village of Umm Al-Khair resembled Susiya, which lay five miles to the south, in that its structures were flimsy, under threat of demolition, and powered by Comet.

Sundown. Across the street and behind an electric fence, in the Jewish settlement of Carmel, with its tiled sidewalks, streetlamps, drip irrigation systems, goldfish ponds, and parking bays, it was the Sabbath. Here in Umm Al-Khair, with its ragged tents and communal outdoor oven of smoldering dung and powdery ash, it was still Ramadan. In both realms, it was a time for reflection and fellowship.

There was a question of whether, at this flashpoint moment of conflict, the girls' father would approve of them venturing to the West Bank
("Dangerous for Israeli citizens") and, by extension, whether Tamar was
a bad mother to hazard their safety. But this, I knew, was the only way her
body could take it. She had to show her children how to love their neighbors. Nina, her eleven-year-old, was reluctant to come. A lonely, sulky,
preteenager with a sprinkling of pimples on her forehead and a lack of
awareness about her own beauty, she looked so much like Tamar had
looked when we first met that I kept mistaking her for her mother. Nina
was an angry girl, as we had been—angry, perhaps, with her parents for
divorcing, and with the planet for being a cruel, ungovernable place. But
her self-consciousness melted when Eid handed her a baby goat with
ears of white silk. And whatever lingering tension there was disappeared
completely when he opened the door to his workshop to show us his
sculptures, and we saw the beautiful thing he had done.

There, lovingly arranged on five metal shelves by the door, sat his
sculptures. There were two bulldozers, a dump truck, and an excavator
modeled after Caterpillars, all painted shiny yellow, plus a Black Hawk
helicopter. Each of them was roughly two feet long, incredibly detailed,
and built precisely to scale out of scrap metal and construction materi-
als wasted by IDF bulldozers in past demolitions. They didn't look like
harmful vehicles—they looked like toys. They had tiny little doors and
gearshifts, motorized treads and blades, and hinged arms, side-view mirrors and lights cut from CDs, driver's seats fashioned from shampoo bottles, control panels with miniature dials. This was the detritus of past
demolitions, the implements of destruction recycled into art—objects of
whimsy, curiosity, and play. Watching Nina and Magali's delight in the
sculptures, I thought how much my own children would love them. I
thought, too, of the unlikelihood of an artist resulting from this terrain—as
unlikely as that rosebush growing in Susiya.

Yet here was Eid, that rare grown-up who hasn't lost the light of play.
He picked up his marvelous CAT 972M. It had taken him six months to
complete. "I saw this kind of bulldozer demolishing Palestinian homes,"
he said. In fact, his brother's demolished home was a heap of twisted
metal and rubble not twenty feet from his workshop, and two months
from this moment, Guy would send me video footage of yet another

demolition in Umm Al Khair, including the community center where kindergarten classes were held. But by that time, Eid would be safely on his way to Berlin, to showcase his creations alongside Ai Weiwei's. For now, he demonstrated how the mechanized blades of the chopper turned, how the arm of the excavator bent. He told us, in his way, how power should be put to use. "The idea of this art didn't just come from the occupation. I wanted to challenge myself to make something with my hands that would work."

Outside Eid's workshop, whether transformed by his handiwork or by the sunset over the South Hebron Hills, Nina started singing "The Sound of Music." Tamar and I knew the lyrics, too, because we'd performed the musical in middle school, so we sang along. It was perhaps a strange inversion to be in the occupied Palestinian territories at a moment of siege, channeling the voices of an Austrian family escaping the Nazis, but it also felt good to sing. Magali ran off to play among the goats with a village girl. The beauty Ahmad and Mahmoud had spoken of finally yielded itself to me. Over the ridgeline and down the bluff, beneath the pale orange bowl of sky, the land was sumptuous and soft. The hills were alive. Judea. The Wilderness. The Promised Land.

I thanked Eid for letting us in. Umm Al Khair: it means "Mother of All Good Things."

2016–2017

PART III

AT THE RISK OF SPOILING DINNER

729 W. 181st St., Washington Heights, Manhattan

Fish Crow, 3750 Broadway, Washington Heights, Manhattan,
muralist: Hitnes

IT WAS ALREADY TOMORROW

I think a lot about what Elad told me in Palestine. "The story should not be one of suffering, but of concrete deeds and actions accumulated over time." Yet it seems to me that the story of struggle is always both. At a time of annihilation, what action to take? How to make the small good deed matter? How to dwell with suffering, and articulate its scale? In such times, the voice must not be individual but polyphonic. Remember in "Wild Geese," what Mary Oliver wrote: *Tell me about despair, yours, and I will tell you mine. / Meanwhile the world goes on.*

Some scientists say the best way to combat climate change is to talk about it among friends and family—to make private anxieties public concerns. For 2019, my New Year's resolution was to do just that, as often as possible, at the risk of spoiling dinner. I would ask about the crisis at parent-association meetings, in classrooms, at conferences, on the subway, in bodegas, at dinner parties, while overseas, and when online; I would break climate silence as a woman of color, as a mother raising Black children in a global city, as a professor at a public university, and as a travel writer—in all of those places, as all of those people. I would force those conversations if I needed to, through the loving performance of this routine task, the taming power of the small. But, it turned out, people wanted to talk about it. Nobody was silent. I listened to their answers. I noticed the echoes. I wrote them all down. Meanwhile, to balance my sorrow, I continued tracking the birds.

JANUARY

TUESDAY, JANUARY 1

At last night's New Year's Eve party, we served hoppin' John. Nim said that when he used to visit relatives in Israel, he could see the Dead Sea from the side of the road, but on his most recent trip, he could not. It was a lengthy walk to reach the water, which is evaporating.

Chris responded that the beaches are eroding in her native Jamaica, most egregiously where the resorts have raked away the seaweed to beautify the shore for tourists.

MONDAY, JANUARY 14

At tonight's dinner party, Marguerite said that in Trinidad, where they find a way to joke about everything, including coups, people aren't laughing about the flooding.

WEDNESDAY, JANUARY 16

On this evening's trip on the boat Walter built, he claimed with enthusiasm that we might extract enough renewable energy from the Gulf Stream via underwater turbines to power the entire East Coast.

Moreover, Walter predicted, with the confidence of a Swiss watch, that when there is more profit to be made in wind, solar, and hydrokinetic energy, "No intelligent businessman will invest another dime in coal. Economic forces will dictate a turnaround in the next ten years," he said.

MONDAY, JANUARY 21

After Hurricane Irma wrecked her home in Key West, Kristina, a triathlete librarian, moved onto a boat and published a dystopian novel titled *Knowing When to Leave*, I learned over lobster tail.

Pinyon Jay, 3668 Broadway, Washington Heights, Manhattan,
muralist: Mary Lacy

FEBRUARY

Tuesday, February 12

We ate vegetable quiche at Ayana and Christina's housewarming party, where Christina described the Vancouver sun through the haze of forest-fire smoke and smog as looking more like the moon.

Monday, February 18

In the basement of Our Saviour's Atonement this afternoon, Pastor John said he's been preaching once a month about climate change, despite his wife's discomfort, and recently traveled to Albany to lobby for the Community and Climate Protection Act.

Saturday, February 23

"When I see those brown compost bins coming to the neighborhood," said a student in Amir's class at City College in Harlem, "it tells me gentrification is here, and our time is running out."

THURSDAY, FEBRUARY 28

"Just between us," Mik said over drinks at Shade Bar in Greenewich Village, "it scares me that white people are becoming afraid of what they might lose. History tells us they gonna get violent."

MARCH

SUNDAY, MARCH 17

On Saint Patrick's Day, Kathy, who'd prepared the traditional corned beef and cabbage, conversed about the guest from the botanical garden in her master gardening class who lectured on shifting growing zones, altering what could be planted in central New Jersey and when.

TUESDAY, MARCH 19

Sheila, who brought weed coquito to the tipsy tea party, said that when people ask her, "What are you Hondurans, and why are you at the border?" she says, "Americans are just future Hondurans."

MONDAY, MARCH 25

Mat recalled vultures in the trees of Sugar Land, Texas, hunting dead animals that had drowned in Hurricane Harvey, during which he'd had difficulty fording flooded streets to reach his mother's nursing home.

APRIL

TUESDAY, APRIL 16

After a bite of roasted-beet salad in the Trask mansion's dining room, Hilary spoke of the historic spring flooding in her home state of Iowa, where the economic impact was projected to reach two billion dollars.

THURSDAY, APRIL 18

Carolyn warned me at the breakfast table, as I picked up my grapefruit spoon, that I may have to get used to an inhaler to be able to breathe in spring going forward, as the pollen count continues to rise with the

Canada Warbler, 3668 Broadway, Washington Heights, Manhattan,
muralist: Andres Alvarez

warming world. My wheezing concerned her, and when she brought me
to urgent care, a sign at the check-in desk advised, "DON'T ASK US FOR
ANTIBIOTICS." Valerie, the doctor who nebulized me with albuterol,
explained that patients were overusing antibiotics in the longer tick season for fear of Lyme.

TUESDAY, APRIL 23

On his second helping of vegetable risotto, Antonius reflected that in
Vietnam, where his parents are from, the rate of migration from the
Mekong Delta, with its sea-spoiled crops, is staggering.

SUNDAY, APRIL 28

Due to Cyclone Fani, Ranjit said he was canceling plans to visit Kerala
and heading straight back to Goa, where he would be available for gigs,
lessons, jam sessions, and meals.

Michael said that beef prices were up after the loss of so much livestock in this spring's midwestern flooding, and so he'd prepared us pork
tacos instead.

MAY

FRIDAY, MAY 3

At the head of the table where we sat eating bagels, Aurash said we won't solve this problem until we obsess over it, as he had obsessed over Michael Jordan and the Lamborghini Countach as a kid.

He added that, just as his parents weren't responsible for the specific reasons they had to leave Afghanistan, in general, the communities most impacted by climate change are least responsible for it.

Balancing an empty plate in his lap, Karthik said that New York City (an archipelago of about forty islands), with all its hubris, should be looking to Sri Lanka, another vulnerable island community, for lessons in resilience.

"We have more in common," he went on, "with the effective stresses of low-lying small-island coastal regions such as the Maldives, the Seychelles, Cape Verde, Malaysia, Hong Kong, and the Caribbean than with a place like Champaign, Illinois—"

"I'm from Champaign!" Pamela interrupted, her mouth full. "It's in a floodplain, too!" she cried. "We're all sitting at this table now."

TUESDAY, MAY 7

"Personally, I'm not that into the future," said Centime, who had a different sense of mortality, having survived two bouts of breast cancer. She uncorked the fourth bottle of wine. We'd gathered at Angie's apartment over Indian takeout for an editorial meeting to comb through submissions to a transnational feminist journal centering women of color. "But I can respect your impulse to document our extinction," Centime allowed.

SUNDAY, MAY 19

Eating a slice of pizza at a kid's birthday party in a noisy arcade, Adam reminisced about the chirping of frogs at dusk in northern Long Island— the soundtrack to his childhood, now silent for a decade.

"Sad to say," he mused, "among the nine million meaningless things I've googled, this wasn't one. It's like a postapocalypse version of my life: 'Well, once the frogs all died, we shoulda known.' Then I strap on a breather and head into a sandstorm to harvest sand fleas for soup."

Roseate Spoonbill, 3531 Broadway, Sugar Hill, Harlem,
muralist: Danielle Mastrion

JUNE

FRIDAY, JUNE 7

Hiral, scoffing at what passes for authentic Punjabi food here in New York, was worried about her family in Gandhinagar and the trees of that green city, where the temperature is hovering around 110 degrees Fahrenheit weeks before monsoons will bring relief.

SUNDAY, JUNE 9

After T-ball practice at Dyckman Fields, while the Golden Tigers ate a snack of clementines and Goldfish crackers, Adelaine's dad, an engineer for the Department of Environmental Protection, spoke uneasily of the added strain upon the sewage system from storms.

SATURDAY, JUNE 15

Jeff, who'd changed his unhealthy eating habits after a heart attack, said, "We are running out of language to describe our devastation of the world."

Lacy agreed, adding, "We need new metaphors and new containers with which to imagine time."

SUNDAY, JUNE 16

Keith confessed that he was seriously losing hope of there being any way out of this death spiral.

TUESDAY, JUNE 18

We sipped rosé, listening to Javier read a poem about bright orange crabs in the roots of the mangrove trees of Estero de Jaltepeque in his native El Salvador, where the legislative assembly had just recognized natural forests as living entities.

The historic move protects the rights of trees, without which our planet cannot support us. Meanwhile, Javier discussed the lack of rights of migrants at the border, recalling the journey he made at age nine, unaccompanied, in a caravan surveilled by helicopters.

In Sudan, where Dalia (who read after Javier) is from, youth in Khartoum wish to restore the ecosystem through reforestation using drones to cast seedpods in the western Darfur region, hoping to stymie disasters such as the huge sandstorms called haboobs.

Owing to this month's massacre, one of Dalia's poems proved too difficult for her to share. "I'd be reading a memorial," she said.

I strained to hear the unspoken rhyme between the rising sandstorms and the dying mangroves, hemispheres apart.

WEDNESDAY, JUNE 19

Salar wrote to me about the call of the watermelon man this morning in Tehran, where groundwater loss, overirrigation, and drought have led to land subsidence. Parts of the capital are sinking, causing fissures, sinkholes, ditches, cracks.

The damage was most evident to him in the southern neighborhood of Yaftabad, by the wells and farmland at the city's edge. There, ruptures in water pipes, walls, and roads have folks fearing the collapse of shoddier buildings. The ground beneath the airport, too, is giving way. Nevertheless, he was bringing me back a souvenir—a bright pink kaffiyeh.

THURSDAY, JUNE 20

"Our airport's sinking, too!" mused Catherine, who'd flown in from San Francisco for this evening of scene readings at the National Arts Club, followed by a wine-and-cheese reception.

FRIDAY, JUNE 21

"It's not true that we're all seated at the same table," argued David, a translator from Guatemala, where erratic weather patterns have made it nearly impossible to grow maize and potatoes.

Drema, David's associate, quoted the poem "Luck," by Langston Hughes:

> *Sometimes a crumb falls*
> *From the tables of joy, . . .*

Then we went out looking for the Korean barbecue truck.

SATURDAY, JUNE 22

"Say what you will about the Mormons," said Paisley, who lives in Utah, "but they've been stockpiling for the end of days for so long that they're better prepared."

SUNDAY, JUNE 23

At the Stone Barns farm, where tiara cabbages, garlic scapes, snow peas, red ace beets, zucchini flowers, and baby lambs were being harvested for the Blue Hill restaurant's summer menu, Laura spoke hopefully of carbon sequestration in the soil.

Edgily, Lisa argued, "There's not a single American living a sustainable lifestyle. Those who come close either are homeless or are spending most of their time growing food and chopping wood."

TUESDAY, JUNE 25

SJ said their car, as well as eight of their neighbors' cars, including a "freaking Escalade," got totaled by a flash flood in the middle of the night in Charleston without warning. "Living in a sea level coastal city is becoming more terrifying by the day," said SJ. They now check the radar before parking.

Thursday, June 27

Magda turned philosophical before returning to Tepoztlán, Mexico. "What is the future of memory and the memory of the future?" she pondered. We were eating raw sugar snap peas, remarkable for their sweetness, out of a clear plastic bag.

Her eyes, too, were startlingly clear. "My daughter's twenty-seven now," she said. "By mid-century, I'll be dead. I can't imagine her future or recall a historical precedent for guidance . . ." Magda lost her thread.

Meanwhile, Roy had been pointing out the slowness of the disaster; not some future apocalypse, but rather our present reality—a world's end we may look to culturally endure with lessons from *Gilgamesh*, the *Aeneid*, the Torah, and the Crow.

Friday, June 28

The other Adam sent word from Pearl River at breakfast: "Today's temps at camp are going to reach 100. It will feel hotter than that. We'll be taking it slower and spending more time in the shade. Don't forget sunscreen, water bottles, and hats; they're critical to keeping your kids safe."

There was no shade at the bus stop in front of the Starbucks on 181st Street. "Why wasn't climate change the center of last night's Democratic presidential debate?" asked Ezra, a rabbi.

"They didn't talk about it at all in 2016," pointed out Rhea's mom, who preferred to see the glass as half-full. "This is progress!" she exclaimed.

Sunday, June 30

Ryan, my father's head nurse on the cardiac unit, feared the hospital was too understaffed to deal with the upswing of heat-induced diseases. Delicately moving the untouched food tray to rearrange the IV tube, he said, "It's hard on the heart."

Western Bluebird and Rufous-Crowned Sparrow, 1614 Amsterdam Ave., West Harlem,
muralist: Shawn Bullen

JULY

TUESDAY, JULY 2

"My homeland may not exist in its current state, a bewildering, terrifying
thought I suffer daily," Tanaïs said of Bangladesh. "Every time I go to the
coast, there's less and less land and now a sprawling refugee camp. Every
visit feels closer to our end."

WEDNESDAY, JULY 3

"Let's lay off the subject tonight, darlin'," suggested Victor as he prepared
the asparagus salad for dinner with Carrie and Andy, who were back in
town for the music festival.

THURSDAY, JULY 4

Holding court over waffles this morning in the stately dining room of the
Black-owned Akwaaba bed and breakfast in Philadelphia, Ulysses, who
works to diversify the US Geological Survey, said, "We need representa-
tion. Earthquakes affect us, too. Volcanoes affect us, too. Climate change
affects us, too."

Charlie stirred the gumbo pot. He said that his girls' public school had closed early this year because its sweltering classrooms lacked air-conditioning to manage the heat wave. "Our seasons are changing," he said, regarding the prolonged summer break.

While Lucy distributed glow necklaces to her little cousins on the Fourth of July, her aunt learned the fireworks display had been canceled by the Anchorage Fire Department, owing to extremely dry weather conditions. Alaska was burning.

Sunday, July 7

Cyrus yanked off his headphones with bewilderment and looked up from his iPad toward his mom. "It says there's a tornado warning," he cried. All through the airport, our cell phones were sounding emergency alarms, warning us to take shelter. A siren sounded.

"Take shelter where?" begged his mother in confusion. She clutched a paper Smashburger bag with a grease spot at the bottom corner. The aircraft was barely visible through the gray wash of rain at the wall of windows rattling with wind.

Nadia, a flight attendant in a smart yellow neck scarf, served us Würfel vom Hahnchenkeulen in Pilzsauce on the delayed seven-hour red-eye from Philly to Frankfurt, on which each passenger's carbon footprint measured 3.4 metric tons.

Monday, July 8

Owing to a huge toxic algae bloom, all twenty-one of the beaches were closed in Mississippi, where Jan was getting ready to start her fellowship, I learned before tonight's dinner at the Abuja Hilton.

Jan ordered a steak, well done, and swallowed a malaria pill with a sip of South African wine. She referred to Joy Harjo's poem "Perhaps the World Ends Here," which starts:

> *The world begins at a kitchen table.*
> *No matter what, we must eat to live.*

WEDNESDAY, JULY 10

Eating chicken suya in the mansion of the chargé d'affaires, Chinelo spoke quietly of the flooding in Kogi state at the confluence of the Niger and Benue rivers.

"Few Nigerians realize," Buchi said, "that the longevity of Boko Haram in the northeast, the banditry in the northwest, and the herder-farmer crises in the north central are a result of rapid desertification and loss of arable land even as the country's population keeps exploding."

THURSDAY, JULY 11

Jide, a confident and fashionable hustler, slipped me a business card claiming his sneaker line was the first innovative, socially conscious, sustainable footwear brand in all of Africa. His enviable red-laced kicks said, "We're going to Mars with a space girl, two cats, and a missionary."

Stacey, a science officer for the Centers for Disease Control and Prevention, was geeking out about the data samples that would help control the spread of vector-borne diseases like yellow fever and dengue when the waiter interrupted her epidemiological account with a red-velvet cake for my forty-third birthday.

"*Nel mezzo del cammin di nostra vita / Mi ritrovai per una selva oscura / Ché la diritta via era smarrita!*" shouted Nicole, my college roommate from half a lifetime ago, before we had kids, before she went blind. We had memorized the opening lines of *The Inferno*, had crushes on the Dante professor, and knew nothing yet of pain. She was the one who said at my baby shower that I should put on my own oxygen mask first.

TUESDAY, JULY 16

Naheed said, "The southwest monsoon is failing in Nagpur. For the first time in history, the municipal corporation will only provide water on alternate days. There will be no water on Wednesday, Friday, nor Sunday in the entire city for two weeks."

Chido told us that in Harare, she was one of the lucky ones on municipal rotation, getting running water five days out of the week, until fecal sludge appeared, typhoid cases cropped up, and the taps were shut off entirely. "They are killing us," she said.

FRIDAY, JULY 19

Kate said the back roads of Salisbury, Vermont, were slippery with the squashed guts and body fluids of the hundreds of thousands of northern leopard frogs—metamorphosing from tadpoles in explosive numbers—run over by cars.

Centime sent a picture of a memorial for Okjökull, the first Icelandic glacier to lose its status as a glacier. "For your time capsule," she offered. The plaque read, "This monument is to acknowledge that we know what is happening and what needs to be done."

Posed as a letter to the future, the message ended, "Only you will know if we did it."

MONDAY, JULY 22

Morgan wasn't the only one to observe that it was the poorer neighborhoods in Brooklyn that had their power cut off in yesterday's rolling blackouts. The powerless scrambled to eat whatever food was in their fridges before it spoiled. Wealthier hoods were just fine.

TUESDAY, JULY 23

"Can you rummage in my mind and take out the fire thoughts and eat them?" asked eight-year-old Geronimo at bedtime. This was the ritual. He felt safer with his anxieties in my stomach than in his brain.

Just back in Los Angeles from an empowering trek to Sicily, where she'd visited the shrine to the Black Madonna despite sizzling temperatures, Nichelle shared her two rules for dealing with the global heat wave: "(1) Drink lots of water. (2) Watch how you talk to me."

WEDNESDAY, JULY 24

Marking the fiftieth anniversary of the moon landing, Pastor John sermonized, "You'd think after seeing the earth from afar, we would do anything to protect this planet, this home. You'd think wrong."

"We've become drunk on the oil and gas poisoning the waters that

Red-Faced Warbler, 601 W. 162nd St., Washington Heights, Manhattan,
muralist: ATM

give us life," he preached. "And we have vomited that drunkenness into
the atmosphere. Truly, the prophet is right," he said, quoting Isaiah 24:4.
*The earth dries up and withers. The world languishes and withers. The
heavens languish with the earth.*

"We have broken the everlasting covenant," reasoned Pastor John.
"Nevertheless, the Bible tells us that God loves this world."

THURSDAY, JULY 25

At last night's "Intimate Dilemmas in the Climate Crisis" gathering at a
software company on Madison Avenue, we were told to write our hopes
and fears for the future on name tags as a silent icebreaker, then to stick
these messages to our chests and walk about the room. Sebastian's sticker
bore only one word: WAR.

Mary Annaïse, who left the event early, said she worried about her
aging mother down South. "I'm the first person in my family born after Jim
Crow. They fought battles so I could live the dreams my mother couldn't.
How can I talk to her about this existential grief of mine when she's already
been through so much?"

"Having one less child reduces one's carbon footprint 64.6 US tons per year," Josephine from Conceivable Future informed us.

"Why is it so easy to police reproductive rights of poor women and so hard to tell the fossil-fuel industry to stop killing us?" asked Jade, a Diné and Tesuque Pueblo activist in New Mexico, whose shade of red lipstick I coveted.

Friday, July 26

Ciarán set down our shepherd's pie and Guinness on a nicked table at Le Chéile. On one of the many drunken crayon drawings taped to the walls of that pub were scrawled these lines from Yeats:

> All changed, changed utterly:
> A terrible beauty is born.

Protesters from Extinction Rebellion Ireland staged a die-in at the natural history museum in Dublin, where Ciarán's family is from, arranging their inert bodies on the floor among the silent, stuffed "Mammals of the World."

Tuesday, July 30

Ari cooked lamb shoulder chops with eggplant and cilantro puree, a family recipe from Yemen, where swarms of desert locusts, whose summer breeding was ramped up by extraordinary rainfall, are invading crops, attacking farms, and eating trees.

Meanwhile, Yemeni villagers are eating the locusts, shared Wajeeh, catching them in scarves at nightfall, eating them with rice in place of vegetables, carting sacks of them to Sanaa and selling them, grilled, near the Great Mosque.

Wednesday, July 31

When Nelly and I chewed khat with Centime in Addis Ababa a decade ago, discussing creation myths at the New Flower Lounge while high as three kites, we never imagined that Ethiopia would plant 350 million trees in one day, as it did today.

Ovenbird, 3607 Broadway, Sugar Hill, Harlem,
muralist: Cern

Eric distributed Wednesday's fruit share under a canopy in Sugar Hill, Harlem. I took note of the Baldwin quote on the back of his sweat-soaked T-shirt when he bent to lift a cantaloupe crate: *The moment we break faith with one another, the sea engulfs us and the light goes out.*

AUGUST

THURSDAY, AUGUST 1

Off the rugged coast of Devon, where Jane grew up picking wild black-berries, the Cloud Appreciation Society gathered to slow down and gaze up at the sky in gratitude and wonder. Nobody spoke of the modeled scenario released by scientists of a cloudless atmosphere.

"In the beginning," said Elizabeth, who lives in Pass Christian, a block from the Mississippi shore, "before they closed the beaches, I saw the death with my own eyes: dead gulf redfish, dead freshwater catfish dumped from the river. Thousands. I saw a dead dolphin in the sand."

FRIDAY, AUGUST 2

"I'm always so pissed at plastic bags and idling cars, but I feel like there's no point in caring anymore," said Shasta upon learning that between yesterday and today, more than twelve billion tons of water will have melted from the Greenland ice sheet.

SATURDAY, AUGUST 3

Meera grew disoriented when she returned to the Houston area to finish packing up the house that her family had left behind and could not sell; it had been languishing on the market for a year, as if cursed.

SUNDAY, AUGUST 4

Because he dearly loved taking his boys camping in the Mojave Desert, Leonard felt depressed about the likely eventual extinction of the otherworldly trees in Joshua Tree National Park.

MONDAY, AUGUST 5

The El Paso shooter's manifesto said, "My whole life I have been preparing for a future that currently doesn't exist. . . . If we can get rid of enough people, then our way of life can become more sustainable."

In her kitchen, Angie nearly burned the platanos frying in oil on her stovetop. "That ecofascist targeted Mexicans," she said, swatting at the smoke with a dish towel. "He called us invaders."

WEDNESDAY, AUGUST 7

"In the Black Forest," wrote Daniel, "there are mainly firs and spruces. Many of them die because it is too dry. We used to have something called land-rain. That was light rain for days. It's gone. When it rains (like now) it feels like an Indian monsoon. What I really want to say to you about Waldersterben (dying forest): Come now, as long as the Black Forest exists."

FRIDAY, AUGUST 9

Claire, a former Colorado farmer, spoke of intensifying forest fires. "The mountains are full of burn scars like this," she said, sharing a shot from a blaze near Breckenridge.

Common Redpoll, Jacob Schiff Playground, PS 192, West Harlem,
muralist: Creative Art Works

"None of us will be able to say later that we didn't know we were doing this to the earth."

THURSDAY, AUGUST 15

Isobel stopped planning our twenty-fifth high school reunion to study the weakening of global ocean circulation and the tanking of the stock market when the Dow dropped eight hundred points today. Back-to-back, she traced with a painted fingernail the lines of the inverted yield curve and the slowing Gulf Stream.

FRIDAY, AUGUST 16

Zulema wasn't surprised when Pacific Gas and Electric went bankrupt from the billions of dollars in liability it faced from two years of raging California wildfires, though it wasn't a downed power line that ignited the Detwiler Fire, which she fled. It was a discharged gun.

On being evacuated from Mariposa for six days by that fire, whose smoke reached Idaho as it burned eighty thousand acres of trees dried into tinder by bark beetles and drought, she said over soup dumplings: "I almost lost my house. It's surrounded by charred forest now. We're like those frogs in the boiling pot."

SUNDAY, AUGUST 18

"The developers don't live here, so they don't care," said Jimmy, the tuxedoed waiter who served me linguini with clam sauce for lunch at Gargiulo's on Coney Island, where the new Ocean Dreams luxury apartment towers are topping out despite sea level rise. "All they care about is making a buck."

MONDAY, AUGUST 19

Manreet said she felt anxious. Yesterday in Delhi, where her sister-in-law lives, the government sounded a flood alert as the Yamuna River swelled to breach its danger mark.

"Punjab, where I come from, means 'the Land of Five Rivers,'" she explained. "It's India's granary. After a severe summer left the fields parched, the brimming rivers are now flooding them. It's worse and worse each year. I feel weirdly resigned."

SEPTEMBER

TUESDAY, SEPTEMBER 3

Although the sky directly above her wasn't blackened by smoke from the burning Amazon rain forest, Graduada Franjinha saw protests along the road to a capoeira competition in Rio. "It's so sad to see how humankind destroys the lungs of the earth that give us breath," she said.

The lesson for survival she'd stenciled on the wall of the Abada Capoeira Bronx studio where Ben was training said, "Ginga means that effortless flair with which Brasilians conduct their lives. It can be seen in the unique way Brasilians move. Observe a samba dancer, a soccer player . . . Ginga is a graceful approach to everyday situations. It's the capacity to improvise solutions and make things easy, even under rough conditions."

Saddened by the loss of twenty-eight wild horses in Pamlico Sound to a mini-tsunami, Chastity remembered seeing them as a kid and swearing to commit them to her forever memory. "You don't see beautiful things like that and question whether there's a higher being," she said. "You just don't."

WEDNESDAY, SEPTEMBER 4

Chaitali said she can't stop thinking about Grand Bahama after learning that 70 percent of it is now underwater. "Where are all those people going to go?" she asked, mystified and horror-struck.

"It is an unprecedented disaster," said Christian, struggling to control his voice. He'd cut his hair since last I saw him. Dorian was still hovering over his birthplace of Grand Bahama. "Natural and unnatural storms reveal how those most vulnerable are disproportionately affected," he said.

FRIDAY, SEPTEMBER 6

At last night's party, Jamilah, a Trini-Nigerian Toronto-based sound artist and former member of the band Abstract Random, took a bite of pastelito and said she'd like to get to the Seychelles before they drop into the Indian Ocean.

SATURDAY, SEPTEMBER 7

"Eat the fucking rich," said Jessica, in reply to a quarterly investment report on how to stay financially stable when the world may be falling apart.

THURSDAY, SEPTEMBER 9

Arwa feared that the plight of 119 Bahamian evacuees thrown off a ferryboat to Florida for being without visas they did not legally need was a sign of climate apartheid.

WEDNESDAY, SEPTEMBER 11

"Ma'am, I *am* the heat," Maurice replied to the woman in New Orleans's Jackson Square who warned him against jogging outdoors because of the heat advisory in effect.

THURSDAY, SEPTEMBER 12

Maya, proud owner of a Chihuahua–pit bull–min pin mix in Montclair, was saddened to learn that nearly three hundred animals had drowned at a Humane Society shelter in Freeport during the hurricane.

Melissa, incensed, asked why they didn't let the animals out of their damned crates.

"Well, if it's any consolation, a shit ton of *people* died, too," argued Sanaa.

"Tons of babies, tons of elderly and infirm people, even perfectly healthy people died, too. More than 2,500 people are still missing, and 70,000 are now homeless.

"Did you not see the videos of people trapped in their attics with the waves crashing over their houses? Y'all sound fucking stupid," Sanaa fumed.

Friday, September 13

"Did you hear the NYC Department of Education approved absences from school for the youth climate strike next Friday?" Elyssa asked during the Shabbat Schmooze, while the children swarmed around a folding table, tearing off hunks of challah and dunking them in Dixie cups of grape juice.

"I'd rather go to school," said Jacob. His dislike of large crowds outweighed his dislike of third grade.

Wednesday, September 18

Amanda, whom I last saw at Raoul's, where we ate steak au poivre and pommes frites, said she had to sell off half the herd on her family's Texas cattle ranch after a drought left the tanks dry, the lake depleted, and the hayfield shriveled.

She mentioned, almost as an aside, that they'd lost half the honeybees in their hives to colony-collapse disorder in the past five years, too.

"Everyone here is linked to someone who works in oil," she said. "It's the center of the damage, and all that industry makes my efforts feel small. Sailing in Galveston Bay after a tanker spill, I wondered if my soaking-wet clothes were flammable."

Thursday, September 19

TaRessa, from Atlanta, said, "I have always loved awakening to birdsong. This year, for the first time, I hear none." A third of North American birds had vanished from the sky in the span of her lifetime.

Friday, September 20

"I'm here to sign out my child for the climate strike," said a dad to Consuelo, the parent coordinator in the main office at Dos Puentes Elementary.

Greater Sage-Grouse, 3920 Broadway, Washington Heights, Manhattan,
artist: George Boorujy

"By the time they're our age, they won't have air to breathe," worried Consuelo. "They'll be wearing those things on their faces—*mascarillas respiratorias.*"

Ben's sign said, "I'm missing science class for this." He was six, in the first grade, and studying varieties of apples, of which he knew there were thousands. He'd also heard that as many as two hundred plant and animal species were going extinct every day.

Shawna told her daughter on the packed A train down to Chambers Street that a teenage girl had done this, had started protesting alone until kids all over the world joined her to tell the grown-ups to do better, had sailed across the ocean to demand it.

Along Worth Street toward Foley Square, the signs said:

SHIT'S ON FIRE, YO
    COMPOST THE RICH
        THIS IS ALL WE HAVE
            I WANT MY KID TO SEE A POLAR BEAR
                SEAS ARE RISING AND SO ARE WE
                    MAKE EARTH GREAT AGAIN
                        SAVE OUR HOME
                          PLEASE HELP

In yellow pinafores, Grannies for Peace sang "The Battle Hymn of the Republic," while a nearby police officer forced a protester to the ground for refusing to move off the crowded street to the sidewalk. "*Shame!*" chanted the massive crowd in lower Manhattan.

"When our leaders act like kids, then we, the kids, will lead!" shouted a gaggle of outraged preteen girls in Catholic-school uniforms. Their voices grew hoarse, though the march had not yet begun.

SATURDAY, SEPTEMBER 21

Humera's Sufi spiritual guide, Fatima, said, "Alhamdulillah! Let's offer a Fatiha for the young generations, who are inheriting a heavy, sad burden left by their predecessors but who are in process of finding their own voice of goodness. This is a movement of consciousness."

THURSDAY, SEPTEMBER 26

"You need to use an AeroChamber that goes over his nose with the pump so he gets all the asthma medicine," La Tonya, the school nurse, instructed me. Her office was full of Black and brown boys like our son, lined up for the first puff of the day.

FRIDAY, SEPTEMBER 27

"The point of the shofar is to wake us up," Reb Ezra said, lifting the ram's horn to his mouth. He blasted it three times with all he had. "Shana tova!" he shouted. The table was dressed for the new year with apples and honey.

"Who shall perish by water and who by fire?" went a line in the Rosh Hashanah service as we were asked to think about atonement. So began the Days of Awe.

SUNDAY, SEPTEMBER 29

Namutebi said at Andrew's memorial service that in the twenty-five years since that picture of the deceased holding his son in Kampala was taken, Uganda has lost 63 percent of its trees.

MONDAY, SEPTEMBER 30

"The Rollerblades are five dollars," said Abby, who sold books, clothes, toys, puzzles, and games she'd outgrown, spread over a blanket on the

Eastern Whip-Poor-Will, 601 W. 161st St.,
Washington Heights, Manhattan,
muralists: Yumi Rodriguez, Candice Flewharty, and Melanie Sokolow

sidewalk leading to the Medieval Festival, to make money to fight climate change.

OCTOBER

TUESDAY, OCTOBER 8
Danielle made risotto in the pressure cooker for dinner tonight in Marin County to feed her ninety-one-year-old grandparents, who are staying over because they lost power in Sonoma as part of the huge, wildfire-driven blackout.

"I'm scared we're going to end up engulfed in flames," said Danielle. "My grandfather keeps asking when the storm is coming, and I keep trying to explain to him that this isn't like a hurricane."

She was curious about how the rest of America sees this—eight hundred thousand people without power as risk mitigation by the energy utility against wildfires during high winds. She asked, "Do they know this is how we live now?"

WEDNESDAY, OCTOBER 9

"We are okay, but it is starting to get smoky, and we are sorry about our friends closer to the fire," Zulema alerted us. The Briceburg Fire was four thousand acres and 10 percent contained. "PG and E will cut power to the northern part of the county," she said.

FRIDAY, OCTOBER 11

"You're going to feel some discomfort," Dr. Marianne warned me at yesterday's annual gynecological checkup. She inserted the speculum. I stared at the wall with a picture of her taken five years prior, on the white peak of Kilimanjaro.

"Are you in pain?" the doctor asked, discomfited by my tears. The glaciers that ring the mountain's higher slopes were evaporating from solid to gas, the wondrous white ice cap that towered above the plains of Tanzania for as long as anyone can remember disappearing before our eyes.

SATURDAY, OCTOBER 12

In the highlands of Tanzania, Kenya, Ethiopia, and Uganda—where my mother-in-law, Damali, is from—the climate is no longer hospitable for growing coffee. Damali will likely serve hot, milky spiced tea at the family gathering she invited us to with a proper note card through the mail.

Baby Kazuki's mother feared her breastmilk had sickened him after she reintroduced eggs into her diet. And she feared for the eight million people ordered to evacuate their homes as Typhoon Hagibis flayed Tokyo, including the house where her father was born.

SUNDAY, OCTOBER 13

In the park this morning, Ana said her Realtor had advised against the offer she wished to make on purchasing her first home through the subsidized Teacher Next Door program. The house she'd fallen in love with was in a flood zone.

TUESDAY, OCTOBER 15

Romy sent us video of the churches in Damour ringing bells before sunrise to warn people of the raging wildfires. "Lebanon is burning," Romy said. "Probably the biggest fire this country has seen. Please send help."

Amaris said, "Mount Lebanon, the refuge of persecuted native minorities and their history in the Middle East, is on fire. For a place that represents holy land for us, I'm not joking when I say I feel my soul has been set aflame."

And then, as if by listing the scorched villages, she could turn them verdant again, she mourned their names: "Mechref, Dibbeyye, Damour, Daqqoun, Kfar Matta, Yahchouh, Mazraat Yachoua, Qournet El Hamra, Baawarta, Al Naameh . . ."

WEDNESDAY, OCTOBER 16

Yahdon, bred in Bed-Stuy, bought his gold Maison Martin Margiela designer sneakers secondhand to stay sustainably fly, he said.

TUESDAY, OCTOBER 22

Amelia posted a picture of the view from her kitchen window in Quito last week. "*Gracias a Dios*, we escaped the fire and the house is still standing!" she said amid nationwide civil unrest, wherein protesters clashed with riot police and a state of emergency was declared.

"Fossil-fuel subsidies were reinstated to stop the protests in Ecuador, a petrostate where the price of an unstable, fossil-fuel-dependent economy is paid by the poor. It's been a tough week," said Amelia, following up with a picture of a chocolate cupcake. "We all need a treat sometimes."

"What's your position on public nudity?" slurred Elliott, my seatmate on this morning's flight to San Francisco. In Melbourne, where he's from, Extinction Rebellion activists had stripped for a nudie parade down Exhibition Street.

THURSDAY, OCTOBER 24

"Are we under the ocean or in the clouds?" asked Geronimo, looking up at the illusion of undulating blue waves made by a trick of laser light and fog machines at tonight's Waterlicht show, both dream landscape and flood.

"Anyone else have their fire go-bag ready just in case?" asked Lizz, who paints wrought iron in San Diego and writes about brujas. Six hundred fires had burned in California in the past three days.

"For me, as a parent, knowing that my ancestors have overcome the brutality of colonialism gives me hope for the future," said Waubgeshig, originally from the Wasauksing First Nation near Parry Sound, Ontario. "My people have seen the end before."

TUESDAY, OCTOBER 29

Salar, just back from Beirut, described a contrast between streets of festering trash and citizens forming a human chain, across sect, at the start of revolution. "It's like we forgot the planet was our house until it grew so dirty we had to wake up," he said.

WEDNESDAY, OCTOBER 30

Felicia, Mark, Dean, Robin, Dara, Kellen, Alexandra, Roxane, Alethea, Susan, David, and Roy all marked themselves safe in Los Angeles during the Getty Fire, which started near I-405 and Getty Center Drive, destroying twelve homes and threatening seven thousand more.

No word as yet on the safety of Samara, Marisa, Nkechi, Josh, Kelela, Anika, or Laila.

THURSDAY, OCTOBER 31

"It's because of global warming," said Geronimo, dressed as a wizard, when his father recalled having to wear a winter coat over Halloween costumes during his own New York City childhood. The jack-o'-lanterns were decaying. It was 71 degrees when we walked to the parade.

NOVEMBER

FRIDAY, NOVEMBER 1

Naheed brought us back a painting of Lord Shiva, the Destroyer, and his wife Parvati, from the Dilli Haat handicraft bazaar in New Delhi, where schools have closed because of the dirty, toxic air.

TUESDAY, NOVEMBER 5

"I feel guilty," said Alejandra, a City College student, at the first Extinction Rebellion meeting held on campus, the same day eleven thousand scientists declared a global climate emergency.

Violet-Green Swallow, 3662 Broadway, Washington Heights, Manhattan,
muralist: Christian Penn

"Is there going to be food at this meeting?" Hector asked, poking his
head in the door of the nearly empty classroom with its mismatched,
broken chairs. Down the hall was a food pantry. "You'd get more students
to act if you offered food," Hector said, then left.

"Our aim is to save humanity from extinction," said Tom, an Iowa
native. He'd volunteered to give the presentation, having joined the protest back in August. The slideshow included a picture of him drenched in
fake blood at the feet of the Wall Street bull.

"This is a decentralized movement. Our nonviolent civil-disobedience
actions are theatrical. We disrupt the status quo by occupying space. This
was my first time getting arrested," Tom said. "You can do this, too."

"Not me," said Cedric, referencing the obstacles to his participation,
as a Black man. "If I get arrested, will it go on my record? Who pays my
bail?"

Valentin, a full-time rebel since graduating with a degree in archi-
tecture, said we could address the criticism of the rebellion as a white

movement that fetishizes arrests at our next house meeting. Demanding divestment, he added, should be on the agenda.

WEDNESDAY, NOVEMBER 6

"Back home in Ontario, the backyard rinks are gone," lamented Michael, the man we met playing solo street hockey in the schoolyard of PS 187. He showed my boy, wobbling on new inline skates, how to balance himself with a hockey stick, how to gracefully sweep the puck across concrete.

SUNDAY, NOVEMBER 10

At Václav's baby shower, Yana, who'd ordered the usual Mediterranean platter, told him to just rip the wrapping paper off the gift. "That's how Americans do it," she said. Václav held up the bibs, booties, and dresses she'd bought for his baby, due in five weeks.

"Is it just me, or does it feel like this is the last baby we will produce?" whispered Renata, depressed by our aging and shrinking department in an age of endless austerity, with several retirements on the horizon yet no new hires. "It feels like *Children of Men*."

MONDAY, NOVEMBER 11

Geronimo climbed into our bed with *The Children's Book of Mythical Beasts and Magical Monsters* open to a page of flood stories, floods delivered by vengeful gods: Utnapishtim, Viracocha, Zeus, Vishnu, and Chalchiuhtlicue.

"'The Mexican goddess of rivers and lakes once flooded the whole world to get rid of all those who are evil, but those who were good were turned into fish and were saved,'" he read. "Will I be saved?"

"You will be safe because we are relatively privileged, not because we are good," I said, torn between wishing to comfort him and wanting to tell him the truth. "Those who are less safe aren't drowning because they are bad but because they are poor."

THURSDAY, NOVEMBER 14

"Samantha's got serious respiratory issues now, too," said her mother, as we waited for the school bus to drop off our kids outside our building, around the corner from a busy bus terminal, in a neighborhood at the

Hooded Oriole and Lawrence's Goldfinch,
3415 Broadway, West Harlem,
muralist: Christian Penn

nexus of three major highways and the most heavily trafficked bridge in the world.

FRIDAY, NOVEMBER 22

"Are we rebels or are we not?" asked Lena, a French international student studying environmental biotechnology. "The best way to make people know the movement is to plan an action and make demands," she said.

SATURDAY, NOVEMBER 23

"Wow, and here I thought it was going to be just another game," said Aaron, class of '98, after student activists from both schools disrupted today's Harvard-Yale football game, rushing the field to demand fossil-fuel divestment. "I guess I should have gone in to bear witness instead of hanging out at the tailgates."

FRIDAY, NOVEMBER 29

Next to me at Kathy's Thanksgiving table sat her eldest son, who'd driven up for the holiday from Virginia, where he said his neighbors in the coal-fields knew their industry was dead and were understandably fearful of the transition into new lines of work.

DECEMBER

SUNDAY, DECEMBER 8

The Ghost of Christmas Present encouraged Ebenezer Scrooge to do the most he could with the time he had left in the Harlem Repertory Theater's opening-night production of *A Christmas Carol*. The last ghost waited in the wings.

MONDAY, DECEMBER 9

Sujatha said it was getting harder to see outside in Sydney, but the failure of state and federal government action was clear: No mitigation policy. No adaptation policy. No energy-transition policy. No response equal to the task of this state of climate emergency.

"I am worried," she said, as ferries, school days, and sports were canceled because of air quality eleven times the hazardous levels. Mike bought air filters for the house, face masks for their two kids. Shaad had asked her, "Will this be the future?"

FRIDAY, DECEMBER 13

The other Ben had been at the UN climate conference in Madrid all week and felt depressed about our chances of getting through this century "if it wasn't for these kids," he said, sharing a picture of teens with eyes drawn on the palms of their upheld hands. "They are watching and awake."

"We're not here for your entertainment. The youth activists are not animals at a zoo to look at and go, 'Awww, now we have hope for the future.' If you want hope for the future, you have to act," said Vega, a Swedish Fridays for Future leader.

WEDNESDAY, DECEMBER 18

"You know it's bad when the sun looks red and there's ash on every windshield," said Sarah, from Sacramento, who could feel it constricting her lungs.

"What's the right balance of hope and despair?" asked the other Laura.

Purple Finch, 130 Hamilton Place, West Harlem,
muralist: Max Kauffman

FRIDAY, DECEMBER 20

In the Netherlands, where Nina just submitted her doctoral dissertation proposal to the University of Amsterdam, the Dutch Supreme Court ruled that the government must protect the human rights of its citizens against climate change by cutting carbon emissions.

"Everyone not from Australia, *I'm begging you*," said Styli in Sydney. She feared international ignorance due to the lack of celebrity and location. "The truth is, our country is burning alive," she said, on the nation's hottest day on record, one day after its prior record.

SUNDAY, DECEMBER 22

"It looks like an alligator's head," said Ben from the back seat on the drive to Nana's for Christmas. "No, a hydra," said Geronimo. Billowing smoke from the towers of the oil refinery and petrochemical plant to the side of the New Jersey Turnpike at Linden took shifting, monstrous shapes.

Monday, December 23

"It's always the women who pick up the mess at the end of the meal," sighed Angie, doing the dishes at the kitchen sink in a pink T-shirt that said, "SIN MUJERES NO HAY REVOLUCIÓN."

Tuesday, December 24

Though it was the third night of Hanukkah, Rebecca was still preoccupied by the Parshas Noach she'd heard weeks before, admonishing her to be like Noah, who organized his life around saving his family despite the part of him that couldn't fathom the flood.

The hardest pill for her to swallow was this: knowing that a single transatlantic flight for one person, one way, is equivalent to commuting by car for an entire year, she now feels flying to Uruguay to see friends and family for the holidays is a kind of violence.

Friday, December 27

Home in Bulawayo for the holidays during Zimbabwe's worst drought of the century, NoViolet described hydropower failure at Kariba Dam. Downstream from Victoria Falls, shrunken to a trickle, the Zambezi River's water flow was too anemic to power the dam's plants, and so, NoViolet said, there was no running water three to four days a week, and power only at night—"a terrible living experience."

"The time of the month can be a nightmare for women and girls. Showers are a luxury. Those who can afford to turn to generators and solar power, but for the poor, it means adapting to a maddening and restricted life," she said.

Saturday, December 28

"Mom!" called Geronimo from the bath. "I can't breathe."

Sunday, December 29

Ben was disturbed by the dioramas on our visit to the American Museum of Natural History. "Who killed all these animals?" he demanded. "Don't they know this is their world, too?"

"I learned to fish at my grandparents' house on the beach, and now my kids enjoy its calm waters," said Trever from Honolulu. "Every year, the ocean inches higher. We will sell the house next year."

MONDAY, DECEMBER 30

From Gomeroi Country, Alison wrote, "Even away from the fires, we saw a mass cockatoo heat kill on the Kamilaroi Highway near Gunnedah. Willy-willy after willy-willy followed us home down that road. I can't find it in me to be reflective about the decade right now. Love to everyone as you survive this, our night."

"The worst part is feeling helpless, held hostage at the whim of an abusive, inconsistent parent who wreaks havoc, then metes out arbitrary punishment in the name of protecting us," said Namwali, from Zambia, about the failing of the hydroelectric company and the failures of those in power. "In a word, capitalism."

TUESDAY, DECEMBER 31

Another New Year's Eve. On distant parts of the planet, it was already tomorrow. The future was there and almost here. We drank prosecco at Angie's party, awaiting the countdown, while thousands of people in the land Down Under fled from the raging bushfires and headed for the beach, prepared to enter the water to save their lives on New Year's Day.

The screen of my phone scrolled orange, red, gray, black—fire, blaze, smoke, ash. A window into hell on earth. I shut it away to be present for the party and the people I loved. Before he kissed me, Victor said, "Here's to a better 2020 for our country and the whole world."

One hundred forty blocks to the south of us, in Times Square, the ball is about to drop.

2019

Bald Eagle, 3623 Broadway, Sugar Hill, Harlem,
muralist: Peter Daverington

WHILE WE ARE STILL HERE

The birds became integral to my orientation in the city when it felt like the world was closing in. I couldn't tell you the exact cross street of the Municipal Credit Union ATM location where I and other city workers withdraw cash, but I could tell you it's right next to the American bald eagle.

One pre-pandemic night, after Angie's book party at the Alianza Dominicana Cultural Center on 166th Street, I took two friends on an impromptu guided walking tour down Broadway to show them the birds. Kamila lit a joint. By the time we reached the oversized mural of the swallow-tailed kite and other birds at the gas station on the corner of 155th, she was high. I expected her and Centime to be as enchanted with the birds as I was, but the more of them we encountered, the more edgy Kamila became. She said she felt like she was in Hitchcock's *The Birds*—being watched, under attack. The number of birds was alarming to her. She felt increasingly surrounded.

The weed could have been making her paranoid, warping her perception. But as we approached the pinyon jay over Manny's Restaurant and Lounge, near 151st, Centime, who had never seen the birds (despite living in Harlem) and was perfectly sober, said she agreed with Kamila—there was something sinister about them. The murals seemed to her like flags planted by an outside entity, laying claim to the neighborhood.

Centime's uneasy feeling had to do with real estate. Habitat loss. She suspected the birds had landed to invite or delight an incoming class of people with more money than the people who'd lived here for a long time: hipsters, white folks. The kind of people who stencil bluebirds on their coffee cups, make noise complaints, and drive up rents. The word she was searching for is "artwashing"—a pattern by which developers identify a neighborhood flourishing with art galleries as having potential for major profit, or in some cases—as with Goldman Properties in Miami, using terms like "urban renewal"—finance public art projects, like Wynwood Walls, as a driver of such transformation.

Closer to home, in Bushwick, Brooklyn, Mi Casa No Es Su Casa is an organization focused on protecting tenants' rights and affordable housing. It has actively protested against artwashing—specifically, against mural collaborations between artists and real estate developers in the Bushwick Collective—for pricing people out of the neighborhood. Would this bird colony contribute to the displacement and ultimate cultural erasure of residents with deep-seated roots? Centime similarly wondered.

I know this is not the intent of the project, which means instead to connect birds to the local community. Yet in the context of the neighborhood, I can understand such suspicions. I also appreciate that the murals, like all art, may suggest multiple meanings. But then Kamila spotted a bird at the CTown and started to laugh. It was the black tern, cartoonishly depicted in whooshy lines and candy-bright colors. A birder would scratch her head at the illogical rendering; it looks nothing like the black tern. The mural is silly and fanciful, like something one of my children would scribble for fun. We observed its transgression. The bird had broken the second dimension to fling its color onto the tie-dyed T-shirt of the oblivious man beside it. It made us indescribably happy. Some works

Black Tern, 3632 Broadway, Sugar Hill, Harlem,
muralist: Laura Ramón Frontelo

of street art can do this for us—spark joy. So can birds. They bend the straight line, startling us out of our ruts with their riotous colors. When we notice them, they show us patterns.

≡

I first met Karen Taylor years ago, at an exhibit of the family photos of her husband, whose Harlem roots extend eight generations. Karen is the founder and executive director of While We Are Still Here. According to its website, the nonprofit organization formed as "a response to the threat that each building's history would be lost and gone forever, partially due to the passing of time, and partially due to 'gentrification,' which is rapidly altering the environment." Before the wrecking ball of displacement demolishes community memory, Karen aims to codify it, while we are still here, so more people can learn how ordinary Black migrants, primarily from the South, sought refuge from domestic terror and wound up transforming Harlem into a global center of intellectual, artistic, and political influence.

While We Are Still Here is dedicated to the preservation of Harlem's Black history, particularly as it played out in two historic buildings on Edgecombe Avenue, 409 (where Karen lives) and 555. Luminary residents of these addresses in the early and mid-twentieth century included W. E. B. Du Bois, Walter White, James Weldon Johnson, Thurgood Marshall, Joe Louis, Paul Robeson, and Count Basie—though Karen is quick to point out that regular folks lived there, too: beauticians, musicians, and gangsters alike. She believes their collective legacy of fighting inequality leaves a blueprint for our current struggles. I believe she is right.

"Many people seem to have no idea that this community was full of involved and evolved people who lived, worked, thrived, and engaged in activities such as seeing to it that their children were educated, curating art shows in church basements and storefronts or on the sidewalk, and coaching Little League teams, as well as offering instruction in the various art forms," began Karen, when I asked her how she felt about the birds. It struck me that she was speaking about our community in the past tense. Her impulse to preserve its history ran parallel to the mural project.

Swallow-Tailed Kite (and others), 575 W. 155th St., Washington Heights, Manhattan, muralist: Lunar New Year

Unlike me, Karen wasn't a fan of the bird project. Normally outspoken, she had difficulty articulating why she felt provoked by it. It wasn't so much that its namesake was "a slave-owning white nationalist," as she'd learned during her time working for *Audubon* magazine, but that this tribute to his legacy was unfolding now, when Black people in Harlem were being displaced. She allowed that the project might be fueled by someone else's idea of beauty. To her, the birds' faces were frightening. "They all look like raptors," she said, referring to the same five-story mural by Lunar New Year that bugged out Kamila. Furthermore, Karen took umbrage with the proprietor of a new café called the Monkey Cup who said she liked the project because it offered kids something beautiful to look at—as though beauty was lacking before the birds showed up.

There's nothing wrong with sounding the alarm that birds are suffering because of climate change, Karen concluded—but if she had the funding, she'd pay artists to commemorate the Lenape people of the region displaced by European expansionism, and the Black people who made Harlem.

≣

Let us imagine the servant ordered down on all fours
In the manner of an ottoman whereupon the boss volume
Of John James Audubon's "Birds of America" can be placed.

Terrance Hayes offers this image of casual violence in his poem, "Ante-
bellum House Party." An absurd degradation—the Black body as a prop.
The problem with the picture is that it's so easy for us to imagine. Easier
than it is to imagine a sky without birds.

Endangered Harlem, 1883, 1885, 1887 Amsterdam, Hamilton Heights, Manhattan,
muralist: Gaia

On Amsterdam and 163rd Streets, ten blocks from Karen's building, is
one of the Audubon Mural Project's boldest works: *Endangered Harlem*.
Painted by street artist Gaia, it takes up the entire facade of an apartment block. This mural includes four species of migratory songbirds:
the black-and-white warbler, magnolia warbler, scarlet tanager, and tree
swallow (passerines, all). In the top right corner, Gaia painted a portrait
of John James Audubon as a young man; in the bottom right, a photo by

Russell Lee taken in the South Side of Chicago in 1941, during the swell of the Second Great Migration; in the bottom left, the hand of James Lancaster, who led the East India Company's first fleet in 1600, resting on a globe.

"I'm grateful to be able to be a part of the Audubon Mural Project and to have had the opportunity to push this photoshop method of arranging history visually," Gaia shares on Audubon's website. "These three patterns of migration run parallel to one another. But the greatest irony of it all is raising ecological awareness whilst the people of Harlem are endangered of significant gentrification."

I felt that irony while shooting the great gray owl, by the artist Key Detail, one of my favorite bird murals. I wanted to document it before it was gone. I stood there a long time in front of the Daliza Pharmacy, trying to frame the shot so as to put a person on an equal plane. A man confronted me. "You gonna buy this building?" He was missing some teeth, looked in rough shape, was maybe unsheltered. He'd known the owner, he said, back in the day. He felt bewildered by the changes, like at the Chipped Cup: a fishbowl full of white people, as he put it. "No," I told him, "I just like the bird." He considered the owl and spat on the sidewalk. "Fucking birds," he cursed.

2021

Great Gray Owl, 3481 Broadway, West Harlem,
muralist: Key Detail

How Much of Love Is Attention? "Wayfinding," St. Nicholas Park, Harlem,
artist: Chloë Bass

HOW DO YOU LIVE WITH DISPLACEMENT?

I kept another diary of the first three months of 2020. In March, the global coronavirus pandemic overtook all our attention. I abandoned my nascent climate activism to homeschool our children under quarantine, even as I understood that the two crises—climate change and COVID-19—weren't in competition. The only action I could maintain was writing down what people in my network said about what they were losing, or stood to lose, from both threats. (Some said that if the same number of people who eventually took the coronavirus seriously took climate change seriously, we might actually save ourselves.)

Climate grief and coronavirus grief feel strikingly parallel. The solutions to both problems rely on collective action and political will. In both cases, and for the same insidious reasons, the poor suffer more. In the United States, our efforts on both fronts were disabled by a reigning power that denied science and valued individual liberty over the common good. In New York City, at the epicenter of America's outbreak, the virus disproportionately attacked Black and brown low-income communities already plagued by environmental health hazards. The zip codes, like ours, with the worst air pollution have also had the highest coronavirus case counts and fatalities. Many of the voices that make up this chorus, like Luz's, come from these communities and must be foregrounded in the climate conversation that has traditionally marginalized us. Climate change is a crisis of communion; of our relationships with one another and with nature. It was my ambition, in gathering our voices, to suggest that the world is as interconnected as it is unjust.

JANUARY

Baird's Sparrow/Brown-Capped Rosy Finch, 3803 Broadway,
Washington Heights, Manhattan,
muralists: Ralph Serrano/Yazmeen Collazo

WEDNESDAY, JANUARY 1

"Happy New Year from Stone Town, Zanzibar," said Centime, "a place of ghosts, if any exists." Rereading the canon of Black studies, she realized that when taking field notes, the main question should be this: "How do you live with displacement?"

SATURDAY, JANUARY 4

"Whatever actions I can take to align with the earth, I must take them," said Daniel, whose drag name is JoMama, from the Bronx.

SUNDAY, JANUARY 5

"I farewelled my beautiful garden a few weeks ago. I've hung on and hung on but there just isn't the water. Farewell herb tea garden, veggies, wildflowers, and carefully curated collection. All those dreams . . ." mourned Pen, in Queensland.

FRIDAY, JANUARY 10

A week after Trever said the sea was creeping toward the door of the house in Punaluu, Oahu, where he'd learned to fish as a boy, his family decided to sell the house, because the road of the shoreline the house sat upon collapsed into the ocean.

SATURDAY, JANUARY 11
"The weather we're getting today in NYC is a reflection of how we treat the world: trash," said Yahdon, from Brooklyn, where it reached 66 degrees during the second week of January.

WEDNESDAY, JANUARY 15
Over chiles rellenos at Posada del Tepozteco, Tim complained of the air quality in Mexico City, where he lives. "The weather has become a bell jar," he put it.

SATURDAY, JANUARY 18
"It's about leaving something that will outlast us—after the people in the archive are gone, after the archivist is gone, after the world changes," said Laura, who spent four years archiving Radio Haiti after surviving the earthquake and is drawn, these days, to reading postapocalyptic literature.

"Preserving this collection assumes there will be a future, that someone will be alive to remember." What Laura remembers from when she regained consciousness, trapped in the rubble with the corpse of her landlady, was the singing.

MONDAY, JANUARY 20
Geronimo's third-grade curriculum at Dos Puentes Elementary was finally starting to confront the issue. He looked up from his reading homework on the endangerment of the monarch: "A Billion Butterflies Have Vanished."

"Why aren't people listening?" he demanded. "Will humans die, too?"

"We have to fight for butterflies and people," I told him. "That's why we gave you a warrior's name." I put the revolution handbook in the tote bag that said "THERE ARE BLACK PEOPLE IN THE FUTURE," kissed him goodnight, and left for the meeting downtown.

MONDAY, JANUARY 27

"You know there's something really wrong when writers join a group," joked Elliott, at the second Writers Rebel meeting. Snacking on Petit Écolier cookies, we imagined new direct-action tactics to protest extinction that didn't feel grand enough, yet were better than nothing.

Andrew, the joyous troublemaker who had edited the book on creative campaigns for social change, said that common pitfalls of fledgling climate activist groups like ours included unspecific aims and ranks that are too white. *Calling the problem a problem is not the same as solving it*, I thought.

WEDNESDAY, JANUARY 29

Sunday was so windy that Officer J. politely demanded that the fifty white rebels who'd gathered to march on the pedestrian path across the George Washington Bridge break the wooden handles off their hand-painted signs, lest they fly into traffic and shatter a windshield.

"Zero emissions!" yelled a cyclist in support as the protesters walked over the bridge alongside the suicide net. They were spaced ten feet apart so that the cars entering NYC could read their signs:

FLOODS
  EPIDEMICS
    MEGA-STORMS
    WILDFIRES
    FAMINE
      MASS EXTINCTION
      TELL THE TRUTH

THURSDAY, JANUARY 30

Leonard said he regretted having to cancel his trip from LA to China because of the virus. At the same time, health and infectious-disease experts were still sounding the alarm they'd been sounding about climate change making the risk of other novel afflictions much more explosive.

Victor and I counted eight people wearing surgical masks on the subway platform at 125th Street while waiting for the A train to carry us back home after date night at Maison Harlem.

FEBRUARY

Pine Siskin, 3898 Broadway, Washington Heights, Manhattan,
muralist: BlusterOne

SATURDAY, FEBRUARY 1

Eneida said over pizza on movie night at our place that her mother, who has dementia, kept getting lost in the Bronx trying to find the airport. She wished to fly back home to Puerto Rico, though two and a half years after Hurricane Maria, her abandoned house in Cabo Rojo had been overrun by termites.

MONDAY, FEBRUARY 3

Sujatha wrote in shock from Sydney about the fallout of the bushfires, the evacuation of her cousins from the outskirts of the city, and the loss of her friend's farm, where all the animals burned to death.

"The scale of the loss is incomprehensible," she said. "We've spent much of the summer indoors with our air filter on. Climate change has never felt this close."

TUESDAY, FEBRUARY 4

"Have you noticed your planet burning lately? Did you know that faith traditions and science both say that we need to dump fossil fuels? Come and find out more with your congressman, Adriano Espaillat, and local faith leaders," began the invitation in my inbox. I sent an RSVP: Yes.

WEDNESDAY, FEBRUARY 5

"He left us indelible instructions with which to clear the way in this here burning world," said Christian, eulogizing Bajan poet Kamau Brathwaite, whose death we mourned today. From "Words Need Love Too":

> *How to make sense*
> *of all this, all this pain, this drought*
> *scramble together vowels jewels that will help*
> *you understand will help you understand these rain.*

THURSDAY, FEBRUARY 13

Ryan said it was T-shirt weather again in Antarctica; the temperature on Seymour Island recently reached 69 degrees.

SATURDAY, FEBRUARY 15

"Bureaucracies are designed to tell you no," said our congressman, the first to speak at the Faith Forum on the Green New Deal. He felt confident that once the resolution became legislation, the deal would pass the House, but not the Senate.

"The biggest problem in the district office is getting bounced around. Problems grow severe. Time does that. Poor people that contribute less to hurting the environment are suffering most."

"My eight- and ten-year-old kids are living in a world that's a degree warmer than when my mother was a child," said Allegra, born a Southern Baptist in Midland, Texas, and now a NASA scientist, who spoke after the congressman.

Five-hundred-year storms have become twenty-five-year storms, Allegra said, recalling the destruction of the Inwood Hill Nature Center by an almost ten-foot storm surge during Hurricane Sandy. She used to take her kids there when they were small.

"The last time Earth was as hot as today, seas were thirty feet higher." Allegra illustrated this fact with a projected slide of the 110-step stair street a mile north at 215th Street, recently rebuilt by the congressman. There was a dark watermark at step fifty.

"Although sea levels are rising twice as fast in New York City as in the rest of the world," Allegra said, "hope is a discipline for survival that we may as well call 'love.' This day. This panel. This community. Our world. We're here to fight for it."

Rivka, a Talmud teacher and founder of the neighborhood Sunrise Movement chapter, spoke next, sharing the story of the firstborn children of Egypt who waged a war against their parents for refusing to free the Israelites, risking their firstborns' lives.

"We're talking about perversion of morality," said Rivka, making an analogy to the present moment. "We're being begged by our children to save their lives. Failing to meet that responsibility should revolt us. They did not ask us to be born.

"What happens when the generation in power fails to live up to the promise to protect our children? It baffles the mind. Society turns upside down and inside out," Rivka said.

Brother Anthony, a Franciscan monk, believes "our denial of the climate crisis is a fear of death. Spiritually, as a nation, we're morbid, insecure, lack control, don't trust our neighbors or other peoples, we don't believe we can change.

"We resort to tyranny and exploitation and deal death to other peoples and to the earth so that we can hold on to our way of life and hold on to our stuff. This is futility, a useless clinging. I advocate for a re-enchantment with life and creation," the monk continued.

"Another way is possible; hence, another world is possible. To convert means literally to turn together. Time is short. I pray we still have enough time to turn together toward enchantment."

SUNDAY, FEBRUARY 16

"Is it in the floodplain if it's in the gray area, or the blue?" asked Victor. We'd visited a Bronx property during an open house, and he was confused about where it fell on the sea level rise map, projected for the 2050s five-hundred-year floodplain, that Allegra had shared the night before.

"If I'm understanding correctly, the house is a hair outside of the gray floodplain zone," said Victor, studying the map on the screen of his phone. "But truly just a hair."

On our way to the open house, I recalled what Rivka had said. "Even though Noach was a believer, and built the ark as God instructed, he didn't really comprehend the catastrophe until he saw the floodwater with his own eyes. Like Noach, we must act as believers, even as we don't really yet comprehend."

"You can't find wood this strong anymore," said Derek, the home inspector, caressing the beams in the basement of the house that may sit in a future floodplain. "They chopped down all the trees with this many rings a long time ago. They don't exist anymore."

MARCH

Ferruginous Hawk, 3551 Broadway, Sugar Hill, Harlem,
muralist: Ben Angotti

At the climate storytelling workshop, Adam, an atmospheric scientist, admitted bedevilment by language when asked by media to attribute particular hurricanes to climate change. Like Spock, he cannot lie and, as a result, sounds craven. "Wimpy" was the word he admitted to.

"Why can't scientists say the simplest thing?" the other Emily, a journalist, demanded to know why Adam could not also speak about extreme weather as a citizen, as a parent, as a person admitting to fear. "Why aren't you more stark?"

"I don't have an answer," Adam deflected. He worried that words like "crisis" and "emergency," when paired with talk of climate, would lead to authoritarian government action.

"Bring your own plastic bag!!! Traiga su proprio bolsa de plastic," announced the sign on a bodega at 181st Street.

MONDAY, MARCH 2

Sophie said she spent three months crocheting a "show your stripes" blanket too large to fit in the photo of it, which featured her napping cat for scale. "Its colors show global average land ocean temperature changes between 1880 and 2019," she said. Crafting it had calmed her terror.

THURSDAY, MARCH 5

"Oh, you can't trust those flood maps," Meera warned over the squash soup I'd made for lunch, after I mentioned the house we'd fallen in love with. "If the city was truly honest about the properties at risk, we'd have another mortgage crisis on our hands."

FRIDAY, MARCH 6

"Headquarters hasn't told us anything yet about the bag ban," said Jason, the sales associate at Party City, where I bought trinkets to fill the goodie bags for our son's seventh birthday party. He corralled a dozen helium-filled balloons into a giant plastic bag.

MONDAY, MARCH 9

"Just finished my apocalypse shopping," said Honorée, whose stockpile included dried fruit and nuts, ground coffee, plant milk, popcorn, olive oil, vegan sausages, unsweetened peanut butter, seventeen gallons of water, and two dozen eggs. "Better safe than sorry."

WEDNESDAY, MARCH 11

"Maybe God is tryna tell us something?" suggested TaRessa, in Atlanta, after satellite space images showed a drastic drop in Chinese air pollution after the novel coronavirus shut down factories there.

THURSDAY, MARCH 12

"I feel scared," said the woman waiting across from us at gate D10 in LaGuardia airport, slightly embarrassed when she caught me watching her wiping down the armrests and the seat of her chair with Lysol. The airport was nearly empty.

FRIDAY, MARCH 13

"We are all scared," Harouna said, at the empty Pain Quotidien on Forty-Fourth Street. Other servers were hastily removing tables and chairs, having just been ordered by corporate to thin the density and decrease the risk.

Over drinks at the Uptown Garrison on Wednesday night, Ayana, who had a painfully sore throat, said she was scared about how to get to her aged mother in Philadelphia in the event of a lockdown, as well as for her mother's health.

Dr. Stephen, an Upper West Side dentist, predicted that when they finally closed the public schools in NYC, the looting and chaos would begin. I wanted to argue, but he was drilling into my tooth. The dentist proudly showed me his stockpile of surgical gloves, face masks, and hand sanitizer. He would only treat patients now after taking their temperature, he said.

The day before the World Health Organization declared the outbreak a pandemic, Paisley, the poet laureate of Utah, said she seriously did not understand anything anymore. "Every day, reading the news, it's like my whole life has prepared me for nothing."

"My middle of the night revelation was that it is a privilege in this country to say you actually have Coronavirus cuz that means u have access to a test that confirmed it. U have the means to get one, wealthy, and or connected," wrote my neighbor Sheila, a die-hard Prince fan.

"All the rest of us who have any of the whole range of symptoms that are basic to a cold or allergies or flu cannot really know what it is cuz we don't have the connections of this devil ass administration and their rich friends," Sheila fumed.

"Truly, fuck this no-universal-healthcare-having country right now," Sheila continued, right before Trump declared a national emergency—one in which his administration claimed not everyone needed to be tested. *"Ese diablo cabrón mentiroso jamás ha sido mi presidente."*

"Is Grandma going to die from coronavirus?" asked Ben, our just-turned seven-year-old, who was helping me mix the filling for a sweet potato pie while Victor tried to reach his mother in the ER.

Miranda said she was pulling her kids out of school because she felt it was the ethical thing to do.

SATURDAY, MARCH 14

"No one knows what'll happen over the coming weeks in Lagos, Nairobi, Karachi, or Kolkata. What's certain is that rich countries and rich classes will focus on saving themselves, to the exclusion of international solidarity and medical aid," said Centime, coughing over mint tea with honey. Her breast cancer was back, and had spread to her spine and lungs.

While ER doctors in northern Italy were warning us in the States to change our behavior to flatten the curve of contagion, Paolo said the social isolation in Turin was making his aged parents lose their minds.

In Bennett Park, Sula's mom, an NYC public school teacher who sat at social distance on the opposite side of the bench from me while our children played soccer, said that she and her colleagues were desperate for the schools to close.

SUNDAY, MARCH 15

In Fort Tryon Park this afternoon, Jake's mom angered him by snatching away his walkie-talkies before he could share them with his friend, because she feared spreading germs.

The other Ben reported from lockdown in Barcelona that "for the last two nights, the entire city has gone out onto their apartment balconies at the same time to applaud for healthcare workers and declare their own vitality and solidarity and stubborn joy."

Leaving the building to stockpile groceries and goods with his anxious wife and toddler, my neighbor Emmanuel, an NYC tour guide, said he was laid off on Friday and planned to apply for unemployment benefits.

"I knew this was coming but I thought we had more time," I said to Victor when the mayor announced school closures.

"Anyone who would like a grab-and-go breakfast may pick it up at 7:45, and lunch at 11:00," said Principal Tori of Dos Puentes Elementary School, via robocall. The schools were provisionally scheduled to reopen April 20. "There is food for anyone who needs it," she stressed, her voice breaking.

"When is the coronavirus going to go away?" asked Ben, at bedtime. "Is it possible that it will never go away?"

Monday, March 16

"What do you do with a national emergency that requires community action, in a country run by white people who not only do not believe in community but have spent all of history trying to destroy yours?" pondered Hafizah, en route back to Brooklyn from LA.

"Watching the spread of this virus has sort of been like getting fitted for new glasses—you start to see a bunch of things more clearly and catch sight of previously overlooked beauty, while also recognizing the ugliness in some of what formerly impressed you," observed Garnette, in Charlottesville, Virginia, who was having difficulty taking deep breaths.

Ayana's octogenarian mother said she had seen some things in her lifetime and knew that in times of chaos, like this, we mourn what we're losing, dwell on loss, and lack the imagination to see that what we're losing may be replaced by something better.

Nelly drove home down the Taconic between "a waning gibbous moon to my right, an eggy sunrise to my left, with Langston Hughes's *I Wonder as I Wander*, him wandering through Haiti, Cuba, the South, Russia, me wondering how these times will change time, how this time we might change."

Tuesday, March 17

From Amsterdam, Nina reported that people in the Netherlands all applauded the first responders and essential service providers tonight at 8:00 p.m.

"Shelter in place?!" cried Alicia, an incredulous elder. "How will they enforce it? This is New York. Folks won't listen. We're not in a prison or a concentration camp. I'll be damned if anybody tells me I can have a visitor or not. Unless you're paying my mortgage, don't you dare."

Katherine visited her parents in New Orleans, where the Black poor had been stranded in Katrina, and stood six feet away in the driveway. "My son tried to hug my mom. When we reminded him he couldn't, she went inside and came back out with one of those long, grabber tool things at the end of which was a piece of paper on which she'd written 'hug, hug, hug.'"

Wednesday, March 18

Briallen, who previously had a large tract of intestines removed, posted a picture of the votive candles in jars arranged on her windowsill. "I shat myself while praying this morning," she said, before heading to a pharmacy in Elmhurst, Queens, to stock up on Depends.

Looking at the red spread of the virus rendered across the world map, Damali, from Uganda, said she was relieved for once that Africa wasn't the seat of calamity.

Thursday, March 19

Adam, who works in fundraising for New York–Presbyterian Hospital, said a rich banker-donor contacted him to ask for the "special" number to call for testing, and that the hospital was looking at the former quarantine islands around Manhattan as pandemic real estate.

"God has given us to one another so we can care for one another," wrote Pastor John—with whom I last spoke in February, at his office in Cornerstone Church where, over weak coffee, he prayed for me in my bewilderment about the climate crisis, and who is now quarantined with the virus.

Friday, March 20

"Fear and panic do not lend themselves to an empowering home birth," cautioned Kimm, our former midwife, regarding increased interest in home birth among pregnant women anxious about going into the hospital during the pandemic.

Emmanuel's wife, Rebecca, a juvenile defense attorney, said she argued her client's case over the phone instead of going to court.

Rivka said, "We canceled seder, which is difficult to overstate what a big deal that is, culturally. It's our son's first Passover ever, so it's sad. My wife and I will be doing seder alone, in quarantine."

Saturday, March 21

My doppelgänger Aysegul, suffering a fever and shortness of breath while quarantined in Paris, said, "All around the city I know what my friends are having for dinner this evening: roast chicken, coconut–butternut squash soup, potato salad, tofu and ginger noodles."

Lamenting the end of the age of dinner parties, Angie said it was also only a matter of time before the drones keep us on lockdown in NYC, as they are doing in France.

Sunday, March 22

Today, in J. Hood Wright Park, Adam said laboring pregnant people could no longer have their partners with them during birth at the hospital he works for, whose towers were visible beyond the pink weeping cherry tree.

"Today we were supposed to have a memorial service for my father, which we canceled a couple of weeks ago when we saw what was coming. . . . I don't know that I'll have the space to take a deep breath and just miss him. But I do miss him," said Sonya.

Monday, March 23

Tony evacuated the city on Friday amid attacks against fellow Asians for causing the so-called Chinese virus.

"I know many of us would love to be lied to. To be reassured 'this will be over soon.' That 'things won't be different.' But the truth is, things have already changed—and will never be the same," said Yahdon.

"If anyone can spare N95 masks for our birth team so we can use them for attending births and keeping ourselves virus-free, we and the families we care for would really appreciate it," said Kimm.

WEDNESDAY, MARCH 25

"Preparing for the days to come to keep the family safe," said Imani, in Brooklyn, who'd sewn fashionable handmade Ghanaian wax-print face masks.

"Eventually, does the whole world go away?" asked Ben, while watching Miyazaki's *My Neighbor Totoro* after a morning of home school. "Like if everyone dies?"

"Yes, viruses spread very quickly in NYC, clearly faster than anywhere else in the United States. But so do arts, trends, ideas, passions, cuisines, and cultures. If you could map those things, they would look exactly like these pandemic maps, but they would be pandemonium maps," said Adam. "And that is why I love living here, and why I consider myself so lucky to make a life with my family here. I love New York."

THURSDAY, MARCH 26

"Seems like almost overnight, the alienating quiet of these Brooklyn streets has been replaced with basically nonstop sirens," said Mik.

"See you on the other side, brother. Stay strong," said Victor to our neighbor, the other other Ben, who gifted us with a carton of eggs and asked us to take in his mail before fleeing the city with his wife and two kids.

SUNDAY, MARCH 29

"What's going on?" asked an alarmed stranger from his parked car across the street on Friday at 7:00 p.m. We were clapping, hooting, and hollering with full lungs from our front stoop, along with the rest of NYC, for the essential workers on the front lines.

On Saturday, Amy from apartment 5, single mother of twins, gave me her keys and said we were welcome to eat what was left in the fridge. Before nightfall, she fled the city for Connecticut in a rental car, afraid of the president's threat to close tristate borders.

Pastor John said, "The weirdest symptom has to be how it wipes out your sense of smell. Drinking coffee now—it's black, but I can't get a whiff of its scent. I've had it for almost ten days. Can't even smell the rain."

APRIL

Yellow-Billed Magpie and American White Pelican, 135th St. and Amsterdam Ave., West Harlem, muralist: ANJL

WEDNESDAY, APRIL 1

A sign in the window of a liquor store in the Bronx said:

> **COVID-19 IS SOME REAL SHIT!**
> **Cover your fucking mouth!**
> **Shut the fuck up! Buy your shit and**
> **leave immediately.**
> **Absolutely NO titty or sock money!**
> **Stand back at least 6 feet, playa.**
> **Store capacity limited to 5**
> **motherfuckers at once.**
> **You cough, you die.**
> **Drink responsibly.**

THURSDAY, APRIL 2

"You could run from Katrina. You can't run from this," said Maurice, in New Orleans, where Mardi Gras may have served as an amplifying petri dish for the virus. Going-home ceremonies with second lines to send off the dead are now banned.

FRIDAY, APRIL 3

"So many folks are looking forward to summer as a break from COVID-19," said Genevieve, who felt economic recovery must be centered on decarbonization. "But I'm terrified the virus won't have abated, and we'll also be faced with the heat waves, fires, and storms that now fill the months July to October."

SATURDAY, APRIL 4

"Yeah, this is how it's usually done, with Black bodies considered expendable. Black lives worthless except in service to commerce or science," TaRessa said. On live TV, two top French doctors had recommended testing coronavirus vaccines on poor Africans.

Centime, who was symptomatic and feeling weak with a collapsed lung, said, "We need to stop fucking around with theory and say that capitalism, with its industrial body and crown of finance, is sovereign; carbon emissions are the sovereign breathing; 'make work' and 'let buy' must be annihilated; there is no survival while the sovereign lives."

It seemed preposterous to refer to her as "dying," while she was still alive.

2020–2021

I Ching Hexagram #9. (*There is still time.*)

PART IV

WE ARE IN IT

Caution, J. Hood Wright Park, Washington Heights, Manhattan

Connecticut Warbler, Nashville Warbler, and Golden-Winged Warbler,
3507 Broadway, West Harlem,
muralist: Shawn Bullen

THE RAMBLE

During lockdown, traffic dried up on the George Washington Bridge. Without the usual car fumes, the air uptown grew impossibly clear. The din of street life died. New Yorkers with money cleared out for their second homes. It was eerily quiet, and empty. Except for the bodegas, all the storefronts closed. For the first time, I was able to spot the three warblers on the shuttered gate of Monarch Cleaners—a small consolation in the midst of chaos. Prior to the pandemic, that store was always open for business. Nobody had reason for dry cleaning now.

Nine of the ten Manhattan zip codes with the most COVID-19 cases were located uptown. The deaths were proportionate to the same demographic vulnerable to environmental racism—that is, the Black and brown poor. Let me tell you how it was: The shriek of ambulances was incessant. The hospital across from the Audubon Ballroom was besieged. The dead were being forklifted into refrigerated trucks.

The murals of remembrance to the birds in my neighborhood were quickly becoming the backdrop to a grief I was yet to figure out how to grasp, much less memorialize: the lopsided COVID deaths among the Black and brown poor in neighborhoods like ours. Bird-watching is a way of bearing witness—of being transported by the beauty in nature. I'm yanked from that reverie, knowing that Blackness is not a beauty that everyone sees; some see danger, and so my watching is always tenuous, provisional, unstable.

Through biological happenstance, the peak of the pandemic overlapped with the peak of the spring migration of birds along the superhighway of the Atlantic flyway. While the city's hospitals ran out of beds, staff, and ventilators, the birds stopped by for berries and seeds to fatten themselves on their northward journey. We could hear them clearly in that uncanny pause, filling the air with song. The plaintive call of the white-throated sparrow: "Sam Peabody . . . Peabody . . . Peabody." The red-winged blackbird: "Cock-a-r-e-e-e." The ospreys' high-pitched "Killy, killy, killy." The chirruping phrase of the scarlet tanager: "Chick-burr." And the soft, sad coo of the mourning dove.

Into this spectacle of avian sound, New York City Audubon board member Christian Cooper stepped out with a pair of binoculars at Central Park's Ramble on the morning of May 25—Memorial Day. A morning looking at birds in what is supposed to be a protected wild could quickly become a dangerous encounter where one's serenity is revoked by prejudice. "I'm calling the cops. I'm gonna tell them there's an African American man threatening my life," Amy Cooper (no relation) threatened after he asked her to leash her dog, as per park rules. New York City Audubon has as its aim to protect birds and the places they need, today and tomorrow. But who will protect those who are watched with suspicion? Who will conserve those of us who need conserving?

Gyrfalcon, 3750 Broadway, Washington Heights, Manhattan,
muralist: Frank Parga

That same day, in Minneapolis, George Floyd was killed by police. The video went viral on Twitter: Officer Derek Chauvin pressing his knee on George Floyd's neck for nine and a half excruciating minutes, even as Floyd repeated, "I can't breathe," and called out for his mother while horrified onlookers begged Chauvin to show mercy. These two news stories intertwined in the zeitgeist of a pandemic already revealing gross inequities in public health. As Black Lives Matter protests erupted across the nation, institutions that had not previously addressed racial justice as crucial to their missions responded with official public antiracism statements, nature conservation organizations among them. Audubon's statement on the incident in Central Park begins, "Black Americans often face terrible daily dangers in outdoor spaces, where they are subjected to unwarranted suspicion, confrontation, and violence. The outdoors—and the joy of birds—should be safe and welcoming for all people."

My God, the peculiar American insanity of this statement of the obvious. How outrageous that it needs saying at all.

"PROTECT BLACK PEOPLE. COVID-19," somebody spray-painted on a mailbox near our building. "ALL MOTHERS WERE SUMMONED WHEN HE CALLED OUT FOR HIS MAMA," somebody else spray-painted on a boarded-up shop window. Mercy, mercy, mercy. At times, it feels our desecration is wholesale. I could describe what an asthma attack sounds like when my children can't breathe, or the gist of the Talk we gave them, or the scene, after a police raid, when my son asked me to stop walking the city, out of fear I might be shot. But Black pain is not so cheap. And Black joy is not so rare. I would rather you know that the day I took the M4 bus and wondered who had tagged the tundra swan, I was on my way to pick up free seedling kits for my children from the Horticultural Society of New York, so that we could grow pea shoots in the living-room window of our apartment, through which we enjoy watching pigeons and the occasional red-tailed hawk. I would rather imagine the bird Chris Cooper was after that day in the Ramble. Maybe the black-throated green warbler, high in the overstory, calling "Zoo-zoo-zee-zoo"?!

2021

Musical Notation, above 5716 Broadway, Kingsbridge, Bronx

SHELTER IN PLACE

Life treads on life, and heart on heart;
We press too close in church and mart
To keep a dream or grave apart.
—ELIZABETH BARRETT BROWNING

Everything I want to say about New York City happened one afternoon in J. Hood Wright Park at the start of the lockdown. Not since the children were toddlers had we spent so much time in this park, five blocks to the south of our building in Washington Heights. We needed public spaces during quarantine. We needed trees, fresh air, and the opportunity to gather, even at a distance. Without the respite, we'd have gone mad in the cells of our three-room apartment, my family and I, in the bewildering spring of 2020 when the boys turned seven and nine.

The governor and the mayor were in some cockfight about what to name the lockdown and who was running the show. The mayor warned us to shelter in place. The governor thought New Yorkers would riot under such oppressive language; instead, we were "on pause." So long as we behaved, the parks would remain open. But to me, pandemic time felt fast-forwarded, not paused, even as the days without childcare were long, because the city and our access to it transformed so quickly: no more trips to the public library or the Museum of Natural History, no more subway rides. When you are forty-three in New York City, raising children, you have already lost the New York that mattered to you at age twenty-three. The loss I am talking about is something else entirely.

Somewhere at the beginning of the curve, my college roommate called from the West Coast to ask if we'd be leaving the city. Where on earth did she think we would go? To my mother's house in New Jersey, she suggested, or to my brother's house down South? I told her, edgily, that we were staying put. We would tough it out, I said, as if New York gave a fuck whether my family personally abandoned it. The truth was, nobody had offered us shelter. Maybe they feared we'd infect them. Had we thought about renting a place in the country for a time? she asked, genuinely concerned—somewhere less congested, at least until the curve flattened? "With what money?" I bristled.

Here's the thing. Before this virus reared its head, we'd sunk all our savings into a down payment on a run-down house in the Bronx, at the end of the 1 line. The plan was to get away from the George Washington Bridge and the nexus of highways whose traffic fumes had given our kids asthma, along with a third of the kids in our zip code. In the midst of this global pandemic, we were somehow going to close on this house, which was purportedly not in a flood zone, fix it up, and move in. Our jobs were here, uptown, where we were raising our boys, though now, overnight, we had begun working remotely from the apartment where we were also teaching home school. We were committed to New York, as married people with children are committed to each other; we were stuck, without free time or space to admit to regret. "I just think it will be hard, psychologically, to be in the city during this crisis," my college roommate said. "So many people are about to die." I resented her in that moment, because I knew she was right and because we really didn't have the option to leave.

One night in mid-March, the super alerted us to shut the windows. He'd heard some story that turned out to be bullshit about helicopters spraying the city with disinfectant. By April, half our building had emptied out—roughly speaking, the white half. Some of the neighbors handed us their keys before they fled, asking us to shove their packages and mail inside their apartment doors. We said we'd see them on the other side, acknowledging the sudden divide between us. The other side was not the cure, since nobody could say with certainty when the vaccine would arrive and this purgatory would end. The other side was anywhere but here, and when we said we'd see them there, we didn't really mean it, because we didn't really believe they were coming back.

Packages started getting stolen from the lobby. The methadone clinic was closed, the local heroin addicts had nowhere to go, the subway trains were rumored to be moving homeless shelters. The line at the food bank snaked around the corner. The cacerolazo grew more frenzied each night at seven, as if we were rattling the bars of our cages instead of cheering for the essential workers. A lady madly banged a stop sign with the handle of her mop. On the night she stopped doing so, I assumed the virus got her. The crematoriums were running around the clock. "This is hard," wailed my friend Drema from her front stoop in Harlem after her father died of COVID. It took her weeks to find a funeral home available to retrieve his body from the morgue because the death business was so swamped. She'd been a child in the 1970s and claimed this New York felt like that one—the Rotten Apple. Muggings were up. She felt unsafe now walking around the block. "Mourning," she told me later, "is not a passive act."

On Easter, I hid plastic candy-filled eggs in the alleyway, near the trash cans. By Mother's Day, I was desperate to escape. I tried, along with the rest of New York City, to foster a dog. I thought a dog would reduce our stress, but all the shelter dogs were taken. It was true: only the biting pit bulls and hopeless medical cases remained. At some point in May, I ransacked empty apartment 34 for children's books after we exhausted our supply. In apartment 61, they'd forgotten to turn off the alarm clock, and I heard it beeping for three months through the wall, even in my dreams.

My children changed with the abrupt closure of their school. They lost their baby teeth, their tempers, their social lives, their Spanish. The older one, I feared, was growing less cute—not to us, of course, but to others in authority who might, under the wrong circumstances, find him menacing. The changes within the parks were equally abrupt: the playgrounds were locked, the basketball hoops removed, the dog runs closed, the police cruising to enforce restrictions. On the surrounding streets of Washington Heights, the immigrant economy vanished overnight. The sidewalk tables of tube socks, knockoff watches, winter hats, and fruit were suddenly gone, and in their place on the empty sidewalks were discarded blue surgical gloves and piles of dog shit. The mom-and-pop shops shuttered, maybe for good; the signs in the windows of the bodegas required face coverings; the traffic on the George Washington Bridge died down. Down in the subway, without the usual volume of human trash to forage, the rats were said to be eating each other. I hazarded the A train, to bring our taxes to our accountant's midtown office, and found the near-empty train to be a time capsule. Its advertisements from the before time made no sense in the new reality. An orchid show at the botanical garden? A college degree from Apex Tech?

As the death toll mounted, the soundscape changed—sirens, constant ambulance sirens, underscored by birdsong. As food insecurity increased, the kids' basketball coach, who had time on his hands with the league shut down, shifted gears to delivering bags of food to hungry families in the hood. They were unable to make rent and doubling up, two families in one-bedrooms—ten, fifteen people, Domingo said. They were growing desperate. He predicted that by June, they'd be driven by hunger to protest out in the street. Somewhere near peak week, when they were forklifting the dead into refrigerated trucks and digging mass graves in the Bronx, an ER doctor at the hospital up the road killed herself, unable to take the strain. Meanwhile, the flowers were blooming in J. Hood Wright, where we dutifully taught the boys to ride their bikes.

On the day I want to tell you about, the weeping cherry tree was beginning to bud over by the boulders. Through its veil of pink blossoms were visible the towers of the hospital where they'd already run out of beds. Ahmaud Arbery had recently been shot on his jog. Breonna Taylor had just been shot in her bed. George Floyd had not yet been suffocated on the street. We did not yet have the data that the virus was disproportionately taking Black and brown lives from the same zip codes plagued by environmental injustice. We had not yet closed on the house. Those things were on the near horizon. On this gorgeous spring day, the park was full of people trying to stretch out, find solace or play, and my boys and I were among them.

The nine-year-old had sent us on a scavenger hunt. *Find a bottle cap. A leaf. A rock. A flower.* I led the seven-year-old toward the cherry tree to pick a bloom. A well-dressed white woman stood beneath its canopy, serenely photographing its branches with the camera of her phone. I thought I understood what she meant to capture: this cherry tree in the time of corona, this unlikely thing of beauty in the midst of the battered city under the shadow of death. As we drew closer, a soccer ball rolled near her feet, nearly kissing her ankle. It belonged to a Dominican kid, brown-skinned, like my own. In a flash, the woman's affect changed from peaceful to revolted. "Get away from me!" she screamed at the boy, though it was the ball that had approached her, not his body. The boy stopped in his tracks. My boy flinched, too, at the sudden violence in the air. Was he confused by her tone or was he already old enough to get it, and all that lay beneath? The boy's mother, who was heavily pregnant, rose from a nearby park bench like an avenging angel.

"Don't you talk to my son like that," she yelled. "He's a child. He's just playing."

"He's too close to me," the white woman claimed. "He's supposed to be six feet away." Her voice was full of panic and blame.

"So move your own damned body six feet. This ain't your house! This the *park*. He's allowed to play."

Above her mask, the white woman looked displeased to have been addressed in this manner, and maybe a little afraid, but reluctant to back down. "You're teaching him to be disrespectful, just like you," she spat.

"Get the fuck outta here, or I'ma run you out," said the mother. She meant it, too. The white woman had the good sense to leave.

Surely that white woman felt vulnerable to infection. The pandemic was making people jumpy. But it was also magnifying the deep-seated racial and class tensions already in place, tensions over real estate exploitation, gentrification, who owns the city, who belongs to what

White-Faced Ibis, Jacob Schiff Playground, W. 136th St., West Harlem, muralist: Creative Art Works

neighborhood, who gets pushed out of neighborhoods, what it means to be a good neighbor, who has a right to move freely in public space, and whose freedom of movement is policed. When the white woman retreated without apology or consideration for these factors, I bet she thought, *these people are animals.* Witnessing that fight with my boy, I absolutely took the mother's side. I felt her righteous anger for all the countless times in their young lives my own kids have already been criminalized, in some cases by their own teachers. There was a lesson for me, in the boldness of her advocacy. Normally, I'm much more polite. But if that white woman, who had no comprehension of her own base knee-jerk disrespect, had not had the good sense to leave the park when she did, I would have joined the campaign to oust her. Quite simply, she did not deserve the cherry tree because she did not understand how to share. There are those for whom New York was only ever a playground, and I feel similarly about them. They never deserved it.

I insisted on the dark city. / I persisted in the dark city. I'm haunted by these two lines of a poem by my neighbor, Sheila Maldonado, who stayed. I recite them, like a mantra. It is June in New York City as I write this. Twenty-one thousand people have died, and the spring has given way to a summer of struggle. Just as Domingo predicted, there are rebellions in the streets, here in New York and all over the world. Home school is now about the art and history of protest. We are in it, marching with the boys. In the middle of Frederick Douglass Boulevard, we paused in a crowd a thousand strong to kneel with our fists raised, and then got up to walk the distance. We are done with sheltering in place. The sirens are less frequent now. The city is in phase one of reopening, on the cusp of phase two. A lot of the storefronts in our hood are still boarded up. It's unclear how the schools will reopen come fall. The ice cream trucks have returned. The tables are back up on the streets, only now the vendors are hawking masks. The neighbors who had the means to flee have not yet returned. Soon, we will move into the house.

2020

Protect Black People, Cabrini Blvd., Washington Heights, Manhattan

Family Quarantine of Love   3/16 - 4/20

Why we do it	what we don't Do	What we can do
— So we don't spread germs	— We won't scare or bully people.	— go to parks
— To stay healthy	— We'll try our best not to touch our faces.	— See friends outside!
— To keep other people healthy.	— go to school	— fasetime with friends and family!
— Because we understand public health better than Trump	— have friends over	— We can call our family and friends!
— Because we're all connected.	— go to friends houses	— go for a drive
— To be kind	— go to places with crowds (museums, Rock n'Jump, restaurants, library))	— We can be patient and forgive each other!
— To be respectful		— dance parties!
— Because the elders we love are precious wonderful people.		— Experiments
		— Art / Drawing
		— Screen time
		— Do Lingo * With Dad and mom
		— family game night
		— hike

Family Quarantine of Love, h/t Carolina DeRobertis, 804 W. 180th St., Apt. 62,
Washington Heights, Manhattan

IN THOSE DARK DAYS

There were headlines about America failing us, the mothers. It was said we'd been set up for failure, that society let us down. Many of us were working full-time while also doing the bulk of the housework, plus managing your remote learning under lockdown. Working moms were "not okay." We were burned out, in despair. We had no childcare, no support from school, from day care, from camp, which, in the best of times, most of us had struggled to afford because we were underpaid.

We wanted, above all, to keep you safe. You needed us. Little ones. Sweethearts. Every five minutes, you needed us: To brush your hair. To make hot cocoa. To kiss your elbow. To resolve your bickering, your tantrum; to change your diaper; to admire your drawing and draw your bath; to witness your dance, your growth; to help you with math; to say it would be okay like we meant it when we had no earthly idea.

Your needs were astonishing. We fretted over our fitness to fill the gaps in your education. We worried about your development, the effects on your brain of social isolation and so much screen time. We were your god and also your slave. We made your birthday as magical as possible under the circumstances. Your smile was the measure of our success.

Our employers did not consider these factors when we underperformed. After we put you to bed, we kept working. We lost sleep. We swallowed our anger. We did the laundry, again. We researched getting you a puppy, knowing we'd be the one to scoop up the poop. For weeks, we forgot to wash our own hair. It was said our brains were damaged. There was nowhere to hide from you, from the repetition of days. Unmasked, we unraveled. We ran out of power and blew up. At times, we resented our partners. We didn't ask our mothers for help because we feared infecting them. The stress was unmanageable, ghastly. We aged.

Our dark thoughts frightened us. If we died, your life would fall apart. In secret, wept.

≣

You mattered more to us than ourselves. We held you close. Some of us recalibrated our worth; were forced to do so by factors beyond our control. Some of us lost or left our jobs. We traveled nowhere but deeper into your gaze. We knew, to your story, we mattered. Your love was astonishing. Little ones! Sweethearts! The searchlights of your clear eyes were turned on us.

You made the shelter we couldn't escape into a magical kingdom. We played with you down on the floor, our mood overtaken by your delight. Dollhouse. Puzzle. Picture book. You lined up all the animals and toy cars. Occasionally, you were kind to your siblings. Your grammar grew. You drew an equation you said was the cure. Your inventions amazed us. We held you close. It has always been so, no matter how dismal: children will find a way to play. You dilated our contracting world. I'm telling you, wild thing, you dissolved the walls.

We taught you life skills. This is how you open a can and count money. This is how to read a clock and wash your hair. This is how to ride your bike and tie your shoes. This is how babies are made and how we bake a chocolate cake. This is the art of protest, how we make a sign that won't blow away in the wind. We let you skip school to check on the neighbors. We answered your hard questions as best we could—even the ones about the afterlife and the end of the world. We cuddled you on the couch.

At bedtime, we performed the rituals that went with tucking you in. We read you the books that were read to us when we were small. We smelled your heads when you slept. We inhaled you like oxygen. In the glow of the night-light, we watched you breathe and suck your thumb. We traded the tiny teeth you lost for shiny coins beneath the pillow.

Looking back, we'll recall your sticky, small hands on our cheeks, how you loved us more than anything, more than a thousand infinities, though you may not remember—in the dark days, you made our bellies hurt with laughter.

2021

PART V

THE THING WITH FEATHERS

Red-Naped Sapsucker, 750A St. Nicholas Ave., Sugar Hill, Harlem,
muralist: BlusterOne

Gray Hawk, 500 W. 135th St., West Harlem,
muralist: Marthalicia Matarrita

TOGETHER, WE WATCH

I reached the corner of 135th Street and Amsterdam to watch Marthalicia Matarrita, in her paint-splattered pants, finish making the gray hawk. It was November, the saddest month. Even from across the street, the mural's lush green background suggested treetops. I saw the finely barred chest pattern, the rounded wings, the hooked beak, and the magnificent thing Marthalicia had done. She had taken the bird out of the sky, to ground in our midst. And she had elevated us to its height. The hawk was about to land: talons reaching sharply for the perch, gaze intent, tail feathers spread for balance.

Marthalicia was slightly rattled. An old man she knew from the church she used to attend on 133rd had just confronted her, demanding to know: "Are you making this for *us* or for *them*?"

That was the wrong question. At the base of Marthalicia's stepladder were $140 worth of spray paint cans. This was her second go at the hawk, which she chose, in part, because it reminded her of her mother's gray hair. In her first version, the bird was already perched on the branch. But too much white space was left on the wall, and she worried it would get tagged over. So she redid it with its wings spread, working from a photograph.

Marthalicia grew up in the neighborhood, in a string of apartments in what she describes as the "Harlem band," stretching from the west side to the east. Sharp life turns moved her family from one place to the next: They were evicted from 137th and Riverside. For a time, they lived on 145th. Her interest in art made her an outlier. "I'm not artistically supported by my family or my community," she said. The sharp turns continued: She went to SUNY New Paltz to study art, having joined the Army Reserve to pay for school, but was yanked from her studies by a call to Kuwait and a pregnancy. Now, age forty-two, she lives with her two children in a basement apartment in the Bronx, without room to make work, except outside. It's what she can afford.

Birds, in Marthalicia's view, are spiritual messengers. She, too, noticed a pattern: Before a friend or family member would die, she'd see a dead pigeon. Or after someone passed, she'd see white feathers and wonder if it was a sign. "When you're from the city, you look down. You don't call attention to yourself, especially as a woman. You keep your guard up. Nobody looks at the sky." But birds, she felt, were trying to tell her something.

Sometimes they'd land near her—crows, hawks, gulls. Previously, all she'd been aware of were the water bugs, roaches, rats, and pigeons. Remembering the big earthquake in Haiti in 2010, when animals were attuned to the vibrations of oncoming disaster before people, she asked, "And *we're* supposed to be more advanced?"

The artist shakes the can of white. Its bead rattles. She holds the printed picture of the gray hawk up against the wing she's made on the wall, using the paper's edge to make short, straight white lines with the spray paint. The mural is beginning to look real. She lingers on the flight feathers. Then she stops. Looks up. Points out the hawk circling above the building at City College where I teach writing. Her face is a picture of awe. She tells me it's the anniversary of her mother's death. The hawk is doing what hawks do. Together, we watch her painting the sky.

2021

Quilt, photograph by Joanne Shima Raboteau, Princeton, N.J.

YOU HAVE BEEN GIVEN

I had forgotten how to dress. My go-to outfit while languishing for ump-teen pandemic months in our cramped apartment was a T-shirt that said, "THERE IS NO PLANET B," and a pair of grayish hand-me-down pants gone baggy in the seat. The events of my fall in 2021 included: Hurricane Ida, when eleven New Yorkers drowned in their basement apartments; the funeral of my beloved father; and Halloween, when one of my kids came down with the Delta variant while trick-or-treating dressed as a zombie cowboy.

Next, I fell ill with the virus. To keep from infecting the rest of our family, I checked my sick son and myself into a city-run COVID hotel near JFK Airport. In that purgatory, our bags were ransacked for guns and drugs. The doors didn't lock. The nursing staff had to check our vitals at regular intervals to be sure we weren't dead. I was too exhausted by the stack of calamities to recognize I was grieving.

Before my father's decline, he was a preeminent scholar of Black reli-gious history. As brilliant as he'd been, I wasn't sure, at the end of his life, that he recognized me. He had died of dementia, like his mother and sister before him. Now, here I was in quarantine, missing him while also working online and trying to get my picky kid to eat the wet fish sticks delivered to our room. I missed the smell of my dad's pipe, the soft texture of his voice, and his handwriting on the index cards it was his habit to use for note-taking. The idea that the contents of his mind had vanished devastated me.

Keep it together, I told myself. My feverish brain rearranged itself in some way that had me quietly repeating the first eighteen lines of *The Canterbury Tales*.

And specially from every shires ende
Of Engelond to Caunterbury they wende,
The hooly blisful martir for to seke,
That hem hath holpen whan that they were seeke.

I was scaring my child. Maybe I reached for Chaucer because my
father had memorized this passage, too, long before I did, and the poetry
made me feel closer to him. Instead of praying to my father directly for
protection, I sweated out the fever while coughing violently, fearing bedbugs and burglary or an internal cytokine storm, pretending equilibrium
for my son's sake.

For some other strange reason, I was offered a Coach bag as a parting
gift upon checkout. It wasn't a knockoff. It was real. I asked a disgruntled
staff member to explain the connection between Coach and the COVID
hotel. "Look, lady," she asked, "do you want the bag or not?" I took the
bag and promptly forgot I owned it. Then, on Thanksgiving, I tripped
and broke my right foot. The podiatrist who dressed my injury in a cast
was named, appropriately enough, Dr. Greif. The only accessory I wore
that season was in the shower—*if* I showered: a plastic bag to protect the
cast. I was a literal and figurative mess.

In early December, an invitation arrived. My friend Ayana and her
wife were planning a Christmas party. Directions to their house upstate
were included. Children were discouraged. Vaccine boosters were
requested. Festive attire was *required.*

After so much isolation, the thought of a party felt almost illicit. I wel-
comed the invitation but doubted I'd make a charming guest. This would
be my first Christmas without my dad.

"What are you going to wear?" asked my friend Angie, who was also
invited.

Ayana was our epicurean friend, a gifted hostess who threw glittering
dinner parties in the salad days before the shutdown sent us off the rails
and out of each other's company. On one of her birthdays, we dined at
the Breslin, in the lobby of the Ace Hotel, where she'd special ordered
a whole roasted suckling pig. How long had it been since we'd gone to a
party indoors? In my case, a year and eight months. Angie, who was a

gifted hostess herself and whose kitchen had been the heart of our former social life, thrived on parties and suffered their loss. She was giddy to attend.

"I have no idea what to wear," I told her. I understood the assignment as a Black woman: I was to come correct, as my grandmother Mabel might have said. Here was an opportunity to indicate with our adornment, comportment, and style that we had overcome. We couldn't show up at Ayana's looking slovenly. My closet was full of fancy things. So was my jewelry box.

Yet, in my lingering brain fog, I couldn't tell what went with what. A once favorite belted brown coat hung too loosely on my frame. The dresses I had previously had tailored to my body and delighted in wearing to readings, weddings, and conferences now looked strangely youthful, as if they belonged to a more colorful self. Nor, despite Ayana's directions, could I easily understand how to get to the party. I'd been to the house before, but as a nondriver. The journey seemed farther now, more arduous to traverse, with multiple limping steps by subway or cab to train or bus, requiring decisions about schedules I struggled to read. What's more, I wasn't sure I remembered how to be fun. *Maybe I shouldn't go*, I thought.

Angie was undaunted by my anomie. She lent me a red silk blouse and a sparkly hair clip; God bless her. Her enthusiasm lifted my spirits, as it had so many times before. She encouraged me to lower my face mask to apply lipstick and, in a boss move, called us an Uber Black to drive us the two hours up the Hudson River Valley in luxury. "I'm fifty," she reasoned. "And you are my date." I'd been living in a scarcity mindset for so many weeks on end that I forgot such decadence was possible. As Toni Morrison writes: "Beauty was not simply something to behold; it was something one could do." I remembered the Coach bag and dug it out. We split the bill and enjoyed the ride.

Ayana and her wife, Christina, were still cooking when we arrived, dressed in aprons and oven mitts, stirring gravy and toasting pecans in pans. The kitchen smelled of roast chicken and potatoes. Their dog wore a bow tie. So much bounty crowded the counter, I didn't know where to put the sweet potato pie and bottle of rich, creamy coquito I'd brought.

In the living room, the other guests sat in their finery listening to Luther Vandross before a roaring fire, swapping stories about real estate, dogs, the plots of books. Their talk was lubricated by wine.

We applauded when our hosts called us to the dining room. The table was worthy of a magazine spread, dressed with fir boughs, a floral centerpiece, and fine china, the napkins folded into origami shapes. On the menu was slow-roasted pernil, garlic string beans, macaroni and cheese, stuffing, rolls, and a big salad with roasted acorn squash. At every setting lay a maroon-colored place card with a name written in silver script. We were eleven at a table meant for ten. I was surprised to find my seat at the head, squeezed right next to Ayana's.

Ayana clinked her glass with her fork, and the room fell silent. She stood up to give a toast expressing gratitude for the vaccine, the food, and the gift of our company, ending by turning her attention toward me. "The reason Emily is seated here, in a place of honor," she announced, "is that her father has recently died."

I focused on a flickering candle and found myself flush with adrenaline, trembling like the flame. I felt stunned and slightly embarrassed. But more than that, I felt seen. I hadn't understood I needed it said. A pause followed, so concentrated with pity, I thought I might cry. Ayana sat beside me and took my hand. I stopped shaking at her touch.

"I'm sorry for your loss," offered a guest at our end whom I'd only just met. Bev was one half of a couple from South Africa, by way of Durham, where they ran an African diasporic book shop. "Why aren't you dressed in mourning clothes?"

"Yes," added Bev's partner, Naledi, seated opposite. "How are we supposed to orient ourselves to your suffering if we don't know you're in mourning?" She placed her napkin in her lap.

Orient themselves to my suffering? I wasn't sophisticated enough to answer. Obviously, this couple had something to teach me. "Tell me," I said in a voice that sounded almost like begging, "how you mourn."

Though Bev and Naledi came from different tribes with different rites around death, they shared their customs around grief. There was cloth worn by the bereaved to signify they were not in their right mind—they were to be treated differently, with more tenderness and, to a degree, to be let off the hook. In some cases, as with the loss of a significant other, the aggrieved might even shave their head. This showed their loss in the baldest way possible.

By the time their hair grew back in, the mourner was in a new phase of mourning, ready to celebrate the rebirth of the person they'd lost into their role as an ancestor. It was often said that when a person died, a seed had fallen. In a ceremony that might occur one year after burial, that ancestor could be asked for protection and guidance. In this way, the dead were not exactly dead. There were certain rituals to perform as the dead made this passage into spirit. For example, the ghost of that ancestor might be presented with a blanket because they would be cold upon reawakening. Or the dead might be buried with a blanket, for a similar reason.

It came as a great comfort to learn this. My father's death was nested among so many other losses, I hadn't yet grasped its particular hold. He didn't die of COVID. He died in devastating increments, swiftened by social distancing: He forgot how to drive, how to read, how to climb stairs. He forgot the names of his children and the names of the ancestors in the framed photos on his wall. Eventually, he forgot to eat. And I had to say goodbye through a mask, spooning his diminished frame in the rented hospital bed at his home.

When he died, we dressed him under a quilt. We laid it over his body in the funeral casket. Somehow, we remembered that we should keep him warm. During my eulogy, I held up the photos of the ancestors, naming them each—because he had made me to know them, and so that his grandchildren would understand that he was become an ancestor, too. We made meaning out of scraps, connected to something deeper.

Here was Ayana, doing the same. She understood how to make space at her table for grief, to let me find a ritual in the sweetness of community. Another word for that knowledge is "grace." Only after that could we make space for hope.

Sadly, *The Canterbury Tales* is unfinished. The pilgrims never even reach their destination. Their return journey isn't written. Not all of the pilgrims who appear in the general prologue get to tell a story. And so the point of the poem becomes the stories told in the collective, the mixing of social classes, and the pluralism behind the enterprise. To my mind, it is about community.

Someone passed the gravy. Someone else opened another bottle of wine. From the far end of the table, I caught flashes of wit. I took a bite of beans. Above us, the chandelier hung like a crown. Angie asked what everyone looked forward to in the new year. We took turns, going around. One guest looked forward to climbing a mountain. Another, to learning how to live alone after a bad breakup. Christina looked forward to the possibility of fostering a child. Ayana looked forward to resuming classes in person at seminary school. Angie looked forward to learning about her son's high school placement and publishing her next book. I looked forward to seeing what would bloom in my garden come spring. Bev and Naledi looked forward to growing their bookshop, which they'd named Rofhiwa. I asked what that word meant, though in retrospect I might have guessed.

"It means," said Bev, gently, "'You have been given.'"

2022

The author's ancestors

Mabel and Albert Jr., circa 1948

GUTBUCKET

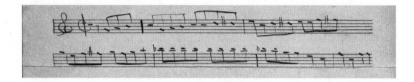

I am a mother raising Black children in New York City, which is unceded Munsee Lenape territory. Often, I am afraid for my children's lives. Where my family lives, the storms are growing worse, and the water is rising. These are not the only threats to our safety. I have come to the Arctic to ask you what changes you have witnessed, and to humbly ask, with your permission, for your wisdom about survival.

This was the script I had rehearsed for my journey to a coastal Alaskan village in the late summer of 2022, when the boys were eleven and nine. It felt potentially extractive and teetering on false equivalence, like something I would have to revise, yet important to get right to justify the carbon footprint of traveling such lengths and leaving my children behind. The last place I'd flown this far from home to study survival was Palestine. I was joining my colleague Dr. Maria Tzortziou, an atmospheric scientist who studies the effects of climate change upon delta systems, including the basin where the Yukon River joins the Bering Sea. This is one of the fastest-warming parts of the planet, or as Maria put it, "the eye of the hurricane." While she prepared to gather water samples with her NASA-funded research team, I would be gathering testimony from elders in the native community whose tribal council office would serve as our base. Increasingly, to win research grants in Indian country, scientific questions needed to derive from the concerns of the people who live there. For instance, *Why are the salmon dying?*

To say I felt like I was traveling to the edge of the earth only indicates my bias. Technically, I wasn't even leaving the United States. To the Yup'ik people who have lived through subsistence hunting and gathering in the Yukon–Kuskokwim Delta for millennia, that region is the center. It took two days and five flights to get there, each plane smaller and later than the one preceding it:

NEW YORK TO CHICAGO

CHICAGO TO ANCHORAGE

ANCHORAGE TO BETHEL

BETHEL TO EMMONAK

EMMONAK TO ALAKANUK

The old man seated directly behind me on the second of these flights suffered from severe memory loss. Every few minutes, for six straight hours, he asked his companion, who I assumed was his suffering wife, where we were going. "Anchorage, Rafael," she repeated. "How many times do I have to tell you?" Three minutes later, he'd ask her again: "Where are we going?" "I already told you," she'd moan. "You're not connecting the dots." Adding to the sense of disorientation (mine) were the pink drifts of fog lit up by the low-hanging sun, not yet set, though it was late at night, because we were so far north.

On the fourth of these flights, a tiny biplane carrying more weight in mail than human cargo, I found myself buckled in next to a woman in her fifties crying like an animal in a trap. She had no companion to ground her. We were about to take off. Her sobs racked her body and filled the aircraft. I knew better than to ask if she was okay, since she so clearly wasn't. Her pain was elemental, naked, pressed up against me, unconscionable to ignore. Not since Drema keening over the loss of her dad to COVID-19 on her front stoop in Harlem had I heard such a sound. The difference then was that we weren't allowed to touch each other to offer comfort. I could freely offer my seatmate comfort through touch now, but I hesitated, thrown off by jet lag, the weight of the moment, the weight of the era. What day was it? Where was I? What to do? What to say? What was this woman's tribe? Why was she mourning? Would she welcome a stranger's intervention or prefer to be left alone? Then I remembered a piece of advice I'd been given by my mother: If there's ever a question of whether to offer kindness to someone in pain, the answer is yes. Do it.

I laid my hand on her heaving shoulder and pressed a tissue into her clenched fist. The engine kicked up. The propellers whirred. The woman, whose name was Edna, turned her wet, pained face to mine and choked, "I just heard my mother died. I'm going to Emmonak to see her. But I'm too late. Why do I have this feeling that I'm all alone? Would it be all right if you hold me?"

I knew exactly how to hold her, because I was a mother. Because I had been held. I stroked the side of her face, her arms, as if she were my baby. While holding her I forgot that my body was in pain. *Ssssshhhh*, I said. The earth below was not solid, was all puddles and ponds. The delta. For the rest of this flight, I now understood, my role was to orient myself to her suffering. We needed a ritual to deliver us. I unwound the bright pink kaffiyeh from my neck and put it around Edna's shoulders. I'd gotten it from my friend Salar. It wasn't a quilt, but it would have to do. "I want you to have this, because your heart is in a dark place," I told her. "This is to remind you that there is still color in the world. When you can see it again, you will give it away to someone else."

"Quiyana," she said, using one corner of the scarf to wipe her eyes. Somehow, I knew the word meant *thanks*. "I really needed the support." Presently, her breath calmed down. We sat together in our travail until landing. Instead of saying goodbye, she asked, "It's never going to be the same again, is it?"

"No," I said. I told Edna I was sorry. I told her I thought I knew how she felt. "I lost my father a year ago. You feel like an orphan."

The singing style of the slaves, which was influenced by their African her-
itage, was characterized by a strong emphasis on call and response, poly-
rhythms, syncopation, ornamentation, slides from one note to another, and
repetition. Other stylistic features included body movement, hand-clapping,
foot-tapping, and heterophony. This African style of song performance could
not be reduced to musical notations, which explains why printed versions do
not capture the peculiar flavor of the slave songs . . .

—Albert Raboteau, *Slave Religion*, page 74

Let me tell you what, in his lifetime, my father loved. He was always
on the lookout for thin places, where the distance between heaven and
earth collapses so we're able to glimpse the divine. My father collected
icons, attempts to render the human face of God. He loved the spiritual
power of the Divine Liturgy. He loved the teachings of Trappist monk,
mystic, theologian, and social activist Thomas Merton, who said: "Life
is this simple: we are living in a world that is absolutely transparent, and
the divine is shining through it all the time." He loved his children. My
brothers. Myself. His grandchildren. My boys.

One area of his research was the retentions, or "Africanisms," that
survived the lacuna of the Middle Passage as expressions of faith among
the enslaved even as they took on, or were made to take on, the master's
religion. The blue note in spirituals. The dance in the ring shout. The syn-
cretism of a Yoruba deity with a Catholic saint. The rituals that protect a
soul's core from being snuffed out by evil forces putting profit above life.
This was the stuff that interested my father.

At his seventy-seventh birthday party, he took down the framed photos from what he called his "ancestor wall." He was well into his diagnosis by then and had started hospice care. His hands were trembling. My
stepmother had baked a German chocolate cake. The pecans in the icing
reminded him of the Mississippi Delta, his home place, in the eastern
floodplain where the Yazoo River joins the Gulf of Mexico—increasingly
a dead zone where the fish are dying, though it's just important for you
to know that this mucky region was also the birthplace of the blues. He
showed us the pictures, my brothers and me and our many children, who
had gathered to celebrate his life, which was ending. As he named our
ancestors for us and reminded us of their stories, I thought—he knows

he is going to forget them. I understood to pay keen attention, lest the names be lost for good.

Agnes Damian, Arthur Chikigak, Donald Augustine, Frank Alstrom Jr., Nellie Edmund, Ragnar Alstrom, Raymond Oney, Regis Augline, Richard Agayar, Rose Isidore, Ruth Oney, Sally Leopold, Mary Ayunerak, Denis and Winifred Sheldon . . .

—List of elders, tribal council office, Alakanuk, August, 2022

The fifth and final flight did not include our luggage, and so Maria and I arrived in Alakanuk with nothing but our questions. Maria was understandably mad: she needed the cooler she'd brought to collect water samples. I was in a state of awe, blown away by the beauty of the wetlands—not water, not earth, but both.

A man called Teddy Hamilton was there at the runway to pick us up. "Welcome to the Yukon," he said. We climbed onto the back of his red four-wheeler. Teddy could talk, smoke, and hunt. He was a self-described "half-breed," and by the time he drove us through low willow thickets along the muddy road to the elevated one-story plywood-sided cabin of his boss, Augusta Edmond, he'd already explained, in as elegant a disquisition on ambiguous loss and traditional ecological knowledge as I'm sure I'll ever hear, that he sees through Yup'ik eyes; that the land is his church with no walls; that he's an outlaw in the eyes of state bureaucrats who try to impose regulations through white man's laws, thinking they know best, when it's local communities who know how to take care of the land sustainably; that his people used to hunt in quadrants in rotation across the domain to keep regeneration in balance, following the seasons and the animals; that in camp, on the tundra, you have to wash town off you and get quiet; that a September moose tastes better than an August moose; that there are more moose now that the willow trees they eat are getting taller, now that there's not enough snow; that the snow geese and cranes are the smartest birds; that there are fewer ducks now in the sloughs than before, and fewer porcupines, too; that you never used to see a beaver when he was a kid, but now that the terrain is different, they're all over the place; that the fat of the black bear turns blue after the thaw, when they grow drunk from eating so many blackberries; that the speck in the sky over there is a loon; that down in the lower forty-eight, where they paved over the wetlands, they think they know better; that the white man's food makes you sick and the white man's ways have poisoned the salmon; that from the headwaters to the mouth, nobody is fishing for salmon because there are no salmon passing through; that things are gonna get harder; that the spring freshet used to feel like an earthquake in his heart when the ice came downriver, but not no more.

He parked in a mud puddle. We were at Augusta's house. Up until recently, Augusta managed the tribes' environmental protection programs, and Teddy was her assistant. But she'd just been promoted to administrative director of the tribal council, the big boss. Like a lot of Bureau of Indian Affairs houses in that village of roughly six hundred, Augusta's was at risk of falling into the river, on account of rapid erosion. And like a lot of working mothers, including me, her house was a mess. She pulled two pairs of sweatpants out of an enormous pile of clean laundry yet to be folded for Maria and me to use as pajamas until our bags turned up. The sweatpants belonged to her daughters. She had five daughters, she told me later, when I asked what she feared and hoped for her children, and though it worried her to imagine—along with the loss of subsistence hunting ways, and the loss of the sea ice, and the loss of the permafrost, and the loss of the salmon—she understood why her girls might not want to stay in the village, since the village would inevitably have to move to firmer ground. But, Augusta also said, she was busy, she had things to do; there were two hundred boxes of meat from the Food Bank of America on the way, they would have to be distributed; so if I really wanted to know about climate change, I should talk to the elders, who remembered how it used to be before the warming began.

Death is someone you see very clearly with eyes in the center of your heart: eyes that see not by reacting to light, but by reacting to a kind of a chill from within the marrow of your own life.

—Thomas Merton, *The Seven Storey Mountain,* page 107

After my father died, a friend introduced me to the term "ambiguous loss." It seems important to mention that this friend was preoccupied by the moral question posed by the climate emergency of bringing children into this world. Ambiguous loss, Meehan explained, was a theory introduced in the 1970s by social scientist Pauline Boss while researching the experience of unresolved grief among families of soldiers gone missing in the Vietnam War. This kind of loss doesn't come with closure or clear understanding. When a loved one goes physically or mentally missing, it leaves the aggrieved unsettled, unresolved, searching for answers. It can occur, as in families scarred by the Holocaust, or slavery, across generations. The therapeutic guidance for such trauma is toward building resilience. Other forms of ambiguous loss might result from incarceration, migration, divorce, displacement, miscarriage, terrorism, addiction, pandemic, climate chaos, or dementia, like my dad's.

Indeed, by his seventy-eighth birthday party, my father had forgotten the names of our forebears. And not only their names: He had also forgotten his own name. The names of his children and grandchildren. How to feed himself. How to dress. How to write. How to read. One of the last things he remembered was the first eighteen lines of *The Canterbury Tales*, which he could mouth along with my recitation, according to something like the muscle memory of that aged ballerina whose Alzheimer's didn't rob her of the choreography when *Swan Lake* was played. *Whan that Aprill with his shoures soote / The droghte of March hath perced to the roote, /And bathed every veyne in swich licour* . . . And then, he forgot that, too.

He was not yet dead, but the mind I had known was gone. Did he know where he was going? "It's okay, Dad," I told him, reintroducing him to the old photos one by one. "This is your mother, Mabel, who raised you to overcome. She brought you north to keep you safe. She went gray young and didn't suffer fools. She took you to church every Sunday. That is you, when you were her little boy, sitting on her lap. This is her husband, your father, Albert, standing between your big sisters, Marlene and Alice. She named you after him. This is her favorite sister, your auntie Emily. You named me after her. This is her father, Edward, whose mother, Mary Lloyd, was enslaved . . ." *These are our people*, I meant to say, *from whom we come and to whom we belong.*

The flexible, improvisational structure of the spirituals gave them the capacity to fit an individual slave's specific experience into the consciousness of the group. One person's sorrow or joy became everyone's, through song. Singing the spirituals was therefore both an intensely personal and vividly communal experience in which an individual received consolation for sorrow and gained a heightening of joy because his experience was shared.

—Albert Raboteau, *Slave Religion*, page 246

On Sunday, I walked from the tribal council to Alakanuk's Catholic church. The mud sucked at my hiking boots. It was the twenty-second Sunday in Ordinary Time. Ordinary time? In two days, I'd overhear Maria reading bad news from a study about "doomed" ice from the rapidly melting Greenland ice sheet that would eventually raise global sea levels no matter what we do now. In three weeks, a typhoon bigger than Texas would gather in the Bering Sea, its remnants flooding the Alaskan coast, washing away fishing camps and houses, sending people running for shelter. Nearby the church was a tide staff for measuring flood events. The high-water mark was eleven and a half feet. The recommended building elevation was a foot higher than that. The church, like the tribal council, was elevated on stilts. Jesus hung on the cross at the altar. Before the cross, wearing green robes and heavy rainboots, stood the deacon, Denis Sheldon. He had a stooped back and a kind face. The acolyte passing out hymnals was his wife, Winifred. One of the hymnals was in Yup'ik. The other hymnal was in English. In a front pew knelt Mary Ayunerak, wearing a flower-print headscarf knotted below her chin. Like Denis and Winifred Sheldon, she was an elder. I knelt beside her. Maybe because it was the start of moose-hunting season, the church was otherwise empty.

The reading was from the Book of Sirach, on humility: how to live within the covenant, faithful to God in the small things. "We know this lesson is true," Denis Sheldon confirmed in his sermon, "because we were already told by our elders to be humble. I'm glad I was born when

I was, because in my childhood I was taught by elders with knowledge from before the missionaries came." On the wall behind the cross hung a fishnet with a harpoon, two eagle-feather dance fans, animal skins, and a traditional drum. Clearly, the cross had been absorbed into a larger cosmology; Jesus was only a small part of the story.

Denis, Winifred, and Mary sang a song in Yup'ik. Toward the end of the mass, I joined my voice with theirs to sing "Go, Be Justice." I remembered this hymn from my Catholic upbringing for this rousing lyric:

Mary Ayunerak, Alakanuk, Alaska

Catch the tyrants in their lies. I liked singing that song with this congregation. When the service was over, I told them why I had come. They agreed to speak with me, back at the tribal council, the following day.

Mary Ayunerak, who has sixteen grandchildren, said that the Yukon didn't use to be that wide. "Right now, our water is dirty," she said, "it's so messed up; it never used to be like that; it wasn't that warm. Used to be kind of cool. Really nice to drink out of, too, from the river. We didn't used to have sandbars down here. A lot of homes that used to be down here are gone due to the eroding. We used to get a lot of fish, wherever we set net, but now, no.

"We used to work together as a community to get food for the winter supply, and whatever we need, but now . . . people hardly ever go fish camping because no fish to cut. We don't go anywhere anymore like we used to a long time ago. Fish are starting to get pus-y, and they have that ugly smell, and so maybe that's why people don't ever hardly go out as a family anymore. We used to be able to go out and get whatever we need and eat together, connect things together. Our homes are sinking. The ground is getting soft. Our permafrost is not like it used to be long ago. We used to travel a lot and put our food that we catch under the ground in a little house-like thing under the ground where we kept our food nice and cool, but the food started to smell."

One of her grandkids, she couldn't remember which, asked her, "Grandma, where do you think we'll move if we have to move?" and she told them, "I don't know, I might not be here, and if I happen to be here, I'll be too senile to know where we are. It will be up to you guys to decide."

Winifred Sheldon picked up Mary Ayunerak's thread on biodiversity loss: "We had some dead fish floating on the Yukon. We're having a hard time getting the fish, and the fish is very important to us. We eat that all winter right now. And right now, it's getting harder. It really is changing a whole lot compared to when we were small. My mom used to sew boots when we were young, with beaver skins for the bottom and spotted seals for the top. I almost learned how to make the boots, but I went to high school almost five hundred miles from here, and then I forgot how to do it. I had to stay with white people, and I felt uncomfortable at times because I was the only Yup'ik person in there, and sometimes I would feel like, 'Why was I born Yup'ik?' when I was staying with them. After two years, I came home and I never went back. I wish I had just stayed home and learned how to do things like my mom did—like sewing, I would have learned, and make those boots for my grandkids or my kids when they were small."

Denis Sheldon shared his real name, Kituuralria (the one who's passing by), before continuing his wife's story about what had been lost. "It was a time when the whole country here was tundra," he said, "and caribou used to come. During the fall time they would herd the caribou toward the river; that was when they killed what they wanted. Most of the time, it was young calves because calves made the best clothing.

"There's been a lot of changes in the sloughs and channels. Beluga whales are not as many. Some seals are not as many as they used to be—like the spotted seals. Right now, the freeze-ups are getting later and later, and breakups are getting to be earlier and earlier. When we were young, our men did a lot of trapping for mink, otters, muskrats, foxes. In those days, when it was colder, the men would be able to cross the Yukon with their dog teams; they could see the frozen breath of the dogs in the air like smoke.

"When I see pictures like that up there," he said, looking up at the ancestor wall of photos of the dead, some of their names passed down and echoed in the list of elders hanging by the phone, such that the tribal council office felt like a genealogy; like reading Leviticus, "they're all people who took care of the land, the water. Like that man to the right towards the window, with red suspenders—Joe Phillip."

"That's my father," explained Winifred Sheldon.

"I learned a lot from him," Denis Sheldon continued. "He would say, show respect to the animals. They would have fish in ceremonies and rites they did, because they believed that everything has a spirit. Yeah, so our people were very spiritual. The one that they respected most is the Great Spirit of the Universe. They always told us, wherever you go, even if you're all alone, you're not alone, because the One-Who-Is-Not-Seen is there with you. They had many names for Him, the Great Spirit. The Almighty. There were so many names they gave Him. So many names. The greatest command was to get to know what He wants us to be."

I told Denis Sheldon that I'd liked the sermon he gave on Sunday about humility and asked if he had been sent to a Catholic residential school as a child.

"Yeah," he said, "close to a hundred miles from home." He was eleven, the same age as my eldest, when he went away to that school, and nineteen when he came back. He lamented, "I really missed my family when I went. Those who took care of us, some of them were not good. They mistreated us. Some of those effects . . . I can see how they affected me. I tried not to show it."

I asked if he found the pope's recent apology for the church's role in the cultural genocide that went down in the state-sponsored residential schools with mottos like, "Kill the Indian in the child," wherein children like Denis and his wife, Winifred, were torn from their families, had their hair shorn, were punished for speaking their native languages, and in many cases were psychologically, spiritually, physically, sexually abused, and even murdered, all in the name of being "civilized"—meaningful. I didn't ask it like that, though. I simply asked if he accepted the pope's apology. He said yes, but also that the apology wasn't enough. In the deacon's view, the expression of sorrow needed to be spread to those who hadn't heard it. "Our children and grandchildren," he said, "I think they need to know that some things happened that were not good. That affected their lives somehow." That was how Denis Sheldon spoke about generational trauma.

"How do you reconcile the need for justice with the need for forgiveness?" I asked him. "How does your body bear it?" This was what I had come all this way to learn—*what do we do with our anger? What can we do when we can't just stop living?*

"That's easy." The deacon smiled. His eyes had been smiling all along. Coming from someone else's mouth, his answer may have sounded trite. But not when Denis Sheldon said it. I felt completely disarmed by his qanruyutet, knowing in my bones these words of wisdom to be true. "We take care of each other."

How do you transform and transfigure sorrow into joy? That's the theme of the greatest Christian poet in the English language of the nineteenth century, Elizabeth Barrett Browning. There's no accident that Du Bois chooses her twice to invoke in the epigraphs of The Souls of Black Folk. *"What is it, Elizabeth?" "I'm trying to saturate Christ's blood in my soul so that I can respond to the ceaseless wailing of suffering humanity and transform sorrow into joy." It's no accident that he would be the one to get inside the humanity of Black folk in the white-supremacist USA. From Bay St. Louis, Mississippi—Gutbucket, Jim Crow Mississippi. Who would think that after the murder of his father, he decided not to hate, not to terrorize, not to traumatize. He's gonna be a love warrior, freedom fighter, yes. That's what he was, but with a deep sense of humility. Deep sense of self-criticism. Oftentimes, I would say, brother, you're too hard on yourself. I'm Holy Ghost Baptist, so I don't have to worry about that. We Baptists don't have*

to worry about being too hard on ourselves. We need to be more humble. Brother Raboteau, going through the rich Catholic tradition and went on his way to Russian Orthodoxy—condemnation of no one, absolute forgiveness of everyone, embracing humanity of everyone but always understanding that you look at the world through the lens of the cross—that's what made my brother a saint in my eyes, 'cause for me, a saint ain't nothing but a sinner who looks at the world through the lens of the heart. And for Christians who look at the world through the lens of the cross.

—Cornel West at Al Raboteau's funeral, September 24, 2021

"And this is your great-great grandmother, Philomena Laneaux," I told my children, seated in the pews at my dad's homegoing, holding up the last of the many pictures I'd brought to the altar—my inheritance. (They will go on the walls of my office now.) I didn't want my children to be afraid of the corpse beneath the quilt, but rather to perceive him as an ancestor, too. The choir waited up in the loft of Mother of God Joy of All Who Sorrow Church, ready to start the next hymn in a minor key. The icons surrounded us, their faces aglow. I felt a sense of communion; an age-old pattern of meaning I rarely had the vision to see; a net of relations stretching behind me into the past and cast out before me into the future. Soon we would gather round the coffin, and then the grave. I finished my eulogy with a quote from Birago Diop's "Sighs," translated from the French, epigraph to *Slave Religion*.

> *Those who are dead are never gone:*
> *they are there in the thickening shadow. . . .*
> *they are in the hut, they are in the crowd,*
> *the dead are not dead.*

2022–2023

His Eye Is on the Sparrow, 135th St., West Harlem, uptown C train platform

THE DREAM HOUSE
AND THE POND

Dissonance
(if you are interested)
leads to discovery.

—William Carlos Williams

I can't talk about our house in the Bronx without telling you first about the pond out front. Given how much worse flooding is elsewhere in New York City—even just two blocks to the east along the valley of Broadway, where the sewer is always at capacity, not to mention elsewhere in the world—I'm embarrassed to gripe about my personal pond. These days, such bodies of water are a dime a dozen. I acknowledge that mine is not the only pond, but merely the pond I can't avoid.

The pond dilates and contracts according to water levels. After a string of dry days, it may shrink to a puddle the size of a serving platter. After a storm, it may stretch the length of five doors end to end, spilling onto the sidewalk over the collapsed curb cut of our driveway on one side and into the middle of the street on the other. The pond is eye-catching in wet weather. It's bad for curb appeal. Its source is environmental, structural, and complex. Its cure doesn't fall easily into one bucket. Except by evaporation, it will not go away. On the rare occasion the pond dissipates, it leaves behind a residue like black mayonnaise.

Because our region is getting wetter with the changing climate, the pond is almost always there. More rain, more storms, more often. The infrastructure of our coastal city, at the edge of the rising sea, isn't fit to handle so much water. Sudden, torrential downpours overwhelm our outdated drainage systems, especially at high tide; drench the subway system and, in some low-lying places not far from our house; turn streets into sewers, basements into life-threatening pools.

In summer, its standing water breeds mosquitoes and collects litter: cigarette butts, scratched-off lotto tickets. Once, I found a used condom floating on its surface. In winter, I worry the pond will become a fall hazard, though since we moved into the house three years ago, the temperature's not been cold enough to freeze it solid. Still, this is what I say when dialing 311 in hopes of remediation. *An elderly neighbor could slip on the ice and break a bone. It could collapse into a sinkhole if left as is.*

"Tell it to the DOT, lady," says the Department of Environmental Protection. I do. "Nope," says the Department of Transportation, "because of the tree, this is a problem for Parks." I follow up. Weeks go by. The Department of Parks winds up passing the buck to the Department of Health. Months go by. "What you need to do for ponding," says the DOH, "is try the DEP." For crying out loud. "I'm being given the runaround," I write to the office of my city councilman. No answer. I follow up. "Reach out to your local community board," I am told. I do. Weeks pass without reply. I follow up. "This wouldn't happen in the rich neighborhood up the hill," I snap. Even the air is cleaner up there. A community associate suggests I contact the city councilman's office. As a city worker myself, I know this dance well—this absurd, disjointed roundelay.

I ruminate over the pond. It has caused me not just embarrassment but distress and shame. I have tried to get rid of it and failed. It has turned me scientific, made me into a water witch. I understand the pond is beyond the scope of any one person, or any one agency, to handle, and that it's perilous to ignore. The pond is a dark mirror. It reflects deeper problems of stewardship and governance and the position of our house in relation to both. The pond is a reservoir of cognitive dissonance. We are privileged to own a home. Yet we live on land that will drown, that is already inundated, in fact. The pond is a portal. Sometimes it smells, this vent hole of the underworld. Beneath its surface, something lies concealed. In the foreboding pond, the house appears upside down, distorted. Given the fact of the pond, why did we buy the house? Now that we dwell in the house, what to do about the pond? How do we live alongside it without it sucking us under?

Technically, the pond isn't on our property at all. Our home inspector had no reason to suspect it. It belongs to the city, along with the road where it lies, and the magnificent curbside London plane tree whose massive roots have rubbled the asphalt, disrupting the flow of rainwater to the catch basin at the corner where our street meets Tibbett Avenue. This is what we were told on the rainy day we arrived to make the final walk-through before closing on the house in the deadly spring of 2020: the pond was up to the city to fix, with taxpayer dollars.

Plenty of folks were deserting New York then. I mean hundreds of thousands. People with more money than us. That we were committed to stay in the city was both an act of necessity and a point of pride. For us, the house was a step up from the crowded three-room apartment in the Heights where we sheltered in place, away from the mad snarl of highways whose traffic gave our boys asthma: a place to stretch out, a sign of our upward mobility—the American Dream. As a Black family without generational wealth, it signified even more: Shelter. Safety. Self-determination. Equity. Arrival. A future for our children. A piece of the pie. All of that. Since some of our ancestors were property themselves, we also believed if we dwelled in the house, we'd be living out their wildest dreams.

We fell in love with the house as soon as we saw it, a run-down detached brick home in a working-class neighborhood with a little garden in back and windows on all four sides, built in 1933. That it was shaded by the biggest tree for blocks felt like good fortune. The house had good bones. Plus, it was a short walk to Van Cortlandt Park, the third-largest park in the city, with a lake and over a thousand green acres to explore. We saw its potential and planned to gut renovate according to our means. We imagined writing books, cooking meals, hanging art, hosting parties, and raising our boys in its sunny rooms. I would plant vegetables and native plants, harvest rain in a barrel, see about government subsidies to finance solar panels. We felt bright about the prospect of homeownership and rejoiced when our offer was accepted. We were moving on up, as in the theme song for *The Jeffersons*. Yet until the day of the final walk-through, we had never visited the house in the rain.

That morning, the pond greeted us like the opposite of a welcome mat, giving shape to whatever latent misgivings we had about making this move. The song shifted key. My heart sank into the gutter. I felt hoodwinked. Behind the grim oval of the pond, beneath the storm clouds, our dream house looked suddenly debunked of its charm, whereas the pond looked substantial enough—as long and wide as a freight car—to merit a name. *Buyer beware!* I waded into the middle of that bad omen to gauge its depth. Murky water sloshed over the tops of my rain boots, drenching my socks. Good Lord. It was so much more significant than a puddle. I wondered what it was, how to name it, and why it was here. I wondered what the land I stood on, with my feet all wet, look like removed of its built environment? Was it actually land, or something less concrete? Could it have been a wetland, once, like Alakanuk? We should have been warned. Why hadn't the pond been disclosed? Because it didn't have to be, said the tight-lipped seller's agent representing the estate of the previous owner, an old man called Jeremiah Breen.

That night, my husband and I lay awake in bed, discussing our options. The alarms of ambulances sounded up from the street. People were dying of the virus all around us. The death toll was climbing. On multiple levels, our survival seemed at stake. We wanted so badly to keep our children safe. Purportedly, the house sat outside the floodplain. But the pond belied our sense of security. Our conversation was tangled, vexed. What if the pond got bigger with worsening weather? Would it pour into the basement? Had we been lied to? Was the house's foundation as solid as we were told? Were we being sold a bill of goods? Could we just suck up the pond with a wet-dry vac? Or was it a sign from above that we should cut our losses and walk away?

We doubted the city would handle the underlying issues—not while hobbled by the pandemic. Were we foolhardy, even though our jobs— jobs we needed for health insurance—were here in the city, to buy a house smack-dab at ground zero of coronavirus contagion, so near the rising sea? What of the future floodplain? Would flood insurance be enough? Would the house be around to bequeath to our children, or would it be underwater? Was it an asset or a millstone? How likely was resale? Would the housing market hold? How high would the waters rise, how soon?

Hadn't I heard someone speculate that the safest bet in this changing climate would be to buy west of I-87? Was it true, as I'd been warned by my friend Meera, that if the FEMA flood maps were up-to-date, development would end, and New York real estate would die? By staying rooted to the city, were we buying into death? Did we even believe, deep down in our souls, in ownership of this kind? Why fake like we or anyone else could own the land? Were we sellouts? Suckers? Settler colonialists profiting from a legacy of genocide and ecocide?

Such questions of capital consumed us deep into the night. One or the other of us recalled Kierkegaard to get us out of our quandary. "Do it, or don't do it—you will regret both." The bottom line was this: if we pulled out of the deal, we'd lose our down payment, amounting to two years of college tuition for one of our kids. By dawn, we admitted our disillusionment. We'd already crossed the Rubicon, imbricated in the twisted system that brought about the pond. Or so we said because, nevertheless, we still loved the house.

We ended up renegotiating the purchase price. Assigning a value to the time it might take to pressure the city to fix the pond was the compromise we thought we could live with. Despite the partnership in our marriage, I understood this invisible labor would likely fall to me, as the wife. And so, I settled. I felt sort of sick at the signing, morally askew, out of my body, floating above my own muddled head. The stack of papers was prodigious, nonsensical, yet legally binding. Now we were anchored by a home mortgage and the worrisome proximity of that home to a body of water that was trying to recall where it belonged.

Every homebuyer has a similar story of entrapment. Every resident in public housing has a hundred stories far worse. Every prisoner in the carceral system has a story about actual entrapment to put this all into perspective. After closing, I burned sage in the attic to rid the place of ghosts. Then I dreamed an ember had dropped from the smudge stick in my hand, burning the whole house down.

Later, I learned that current maps for flood risk often overlap with maps of historic housing discrimination. Geography determines a neighborhood's risk and, this being America, so does race. Neighborhoods that suffered from redlining in the 1930s—when our house was built—face a far higher risk of flooding today. Appraisers mapped cities in the New Deal era, assigning grades to neighborhoods based on several factors, race chief among them. Black and immigrant neighborhoods like ours were deemed undesirable, marked by yellow or red lines and assigned poor grades of C or D, designating these areas "declining" or "hazardous" (and, by extension, the racial minorities who lived there as unfit for mortgage lending), spurring decades of disinvestment by which the old divisions are maintained. These redlined neighborhoods suffer far higher risk of flooding today across dozens of major segregated metro areas in the United States, most notably Boston, Sacramento, Chicago, Detroit, and New York. I didn't know all this when we bought our house in the Bronx, but my subconscious did. The pond suggested a submerged history beneath the daily surface of things.

The renovation was costly and overlong, as renovations always are—in our case delayed by labor shortages and disruptions to the global supply chain. As the months ticked by, we were forking out mortgage on top of rent, hand over fist. When our lease ran out, the house still wasn't ready. We sublet the ground-floor apartment of a concert pianist who'd returned to Russia when the shit hit the fan, furnished with a grand piano we were forbidden to play and infested with bedbugs. The work went on and on.

The house was not just a risk but a wreck. Its rusty tanks sweated out oil that looked like blood onto the basement floor. The bulk of its windowpanes were cracked; its floors, uneven; its doors, out of plumb. It lacked adequate insulation. Despite the sage, it was obviously haunted, this money pit. Impatiently, through sweat equity, by the grace of God and a good guy called Bruno, we fixed it up. Bruno promised we'd know we were finally home when we hung our pictures on the walls. We tore the walls down to the studs. We removed the asbestos. We reglazed the tub, replaced the windows, redid the wiring, and retiled the roof. Bruno found out the upstairs was missing a subfloor. He recommended we lay one down for even footing beneath new floorboards, though this project was way beyond our budget. The price of wood was sky-high. We slunk further into debt.

Under the creaky old planks, we discovered a newspaper dating back to the Depression. The front page addressed the use of antiques in home decoration. It featured a photo of a cardroom with an eighteenth-century Queen Anne table being used for bridge. Seeing the photo made me dizzy, just as I felt when I surveyed the pond, as if I was slipping down a worm-hole, backward in time. How far back could I imagine? The paper flaked into pieces like the wings of brown moths when I tried to turn the page.

By the time Jeremiah Breen took possession of the house, bridge had

"Antiques and Decoration in the Home," *New York American,*
Saturday, September 26, 1931

fallen out of fashion. Bridge was a game his parents might have played
in the evening with friends, at an antique card table like that, in a front
parlor like ours. At the time the table was carved, this part of the Bronx
was marsh. When I input our zip code into the online archive of the US
Geological Survey, I can see on a century-old map what this wetland
looked like before it was developed into the grid of streets, shops, houses,
schools, and apartment buildings that make up the neighborhood now.
In 1900, the land is still veined by blue streams. A pin in the shape of
a teardrop marks the future spot of our present address, amid associated wetland meadows and swamps, smack-dab in a bend of a waterway called Tibbetts Brook. The brook was named after a settler whose
descendants were driven off the land for their royalist sympathies during
the Revolutionary War. Before that, it had another name. The Munsee
Lenape called it Mosholu. We live on the ghost of this rivulet, just one of
the city's dozens of lost streams.

It is difficult to say how I felt when I saw that teardrop. The teardrop
confirmed what I sensed about the true nature of my pond, which was
so much more than a puddle, and not mine at all, but rather a part of a
much larger body of water. I felt many things mixed up together all at
once: alarmed, anxious, defrauded, curious, humbled, and beguiled.

Waterways like Tibbetts Brook were once the lifeblood of the city. I know about this history from my friend Liz's ex-boyfriend Steve Duncan, who put on hip waders and a headlamp and went spelunking in the culverts of the sewers to help unearth it. As New York grew, in the seventeenth and eighteenth centuries, into the world's supreme port, it relied on such freshwater streams for transportation, drinking water, fishing, and waterpower for grain mills and sawmills. As the city developed, the brook became polluted and insufficient for the growing populace along its banks. In the nineteenth century, modern industrialization brought railroad lines that overtook waterways like Tibbetts Brook as transportation routes. Waterpower was replaced by steam. Steam was replaced by electric power. These lifeways became industrial wastelands became Black and brown neighborhoods. Plundered water bodies. Plundered peoples. Devalued lands are where devalued people are made to live.

The works of landscape ecologist Eric Sanderson and scholar of the Lenape Herbert Kraft help me to imagine a preindustrial, pre-European vision of my home place. The Wiechquaeseck community of Lenape lived in a settlement nearby around Spuytin Duyvil Creek, fed by the waters of Mosholu. They lived mostly out of doors and owned only as much as they could carry, working together and relying on each other for survival. Wealth was not a thing to collect or store, in their view. Wealth was in communion with each other, and in balance with the abundant natural world, "filled with an almost infinite variety of plants, animals, insects, clouds and stones, each of which possessed spirits no less important than those of human beings," according to Kraft.

It is not so hard to see a remaining pocket of that natural world that was once my home, still teeming with life. All I have to do is walk three blocks east to Van Cortlandt Park where a narrow belt of lowland swamp forest still survives along a trail around open water. According to the NYC Department of Parks and Recreation, "Though small, this freshwater wetland is ecologically valuable, providing a home for many plant and animal species." It's an ecosystem that slows erosion, prevents flooding by retaining storm waters, filters and decomposes pollutants, and slows global warming by converting carbon dioxide into oxygen.

Pin oak and red maple stand above Solomon's seal, Virginia creeper, marsh fern, and sensitive fern. Hunting the swamp are barred owls and red-tailed hawks. Water lilies, swamp loosestrife, and arrowhead each grow at different water depths, almost choking closed the open water by midsummer. Mallards and wood ducks feed, nest, preen, and glide among dense strands of cattail, buttonbush, arrow arum, and blue flag. Great and snowy egrets and green herons stalk the muddy water, spearing frogs and fish with their beaks. Eastern kingbirds and belted kingfishers screech from the treetops while painted turtles sun themselves on the lodges of muskrats. These, too, are my neighbors.

The Van Cortlandt Swamp is fed by Tibbetts Brook, before the brook divides down into the concrete conduit, its tail buried. This little swamp that remains near my home is a patch of only 2,000 acres of freshwater wetland remaining in the city today out of the 224,000 acres it boasted 200 years ago. Like Joni Mitchell sang, "They paved paradise to put up a parking lot." Many species that once lived in these wetlands have been lost forever.

Remember what Toni Morrison wrote: "All water has a perfect memory and is forever trying to get back where it was." From that point of view, the pond in front of our house is not a nuisance but, rather, the brook remembering itself. Mosholu. It means what it sounds like it means: the small, smooth stones the brook flowed over. How might Thoreau have described my pond? The pond is a gift to the birds who stop there to bathe. The pond is a refuge. In the summer, it is respite from the heat, a cool drink of water riffled by wind. Though not pristinely wild, the pond is a place for wildlife to slake their thirst at night: possum, coyote, skunk. The pond is a lieu de mémoire, a reservoir. When the sun hits it at the right angle, the pond's surface dances with jewels of light. When night comes, the pond throws back the orange glow of the streetlight. When the raindrops plunk into the pond, they cause overlapping, ever widening circles. Morrison again: "Our ancestors are an ever-widening circle of hope."

The pond is the paved-over wetland, reasserting its form. It transcends the mirage of the house. It cleanses perception of artifice and illusion. It catches a spillover of pleasure. The toddlers who attend the home day care run by our next-door neighbor, Margaret, love to stomp and splash in the pond when their parents come to fetch them at the end of the workday. My children would have jumped in the pond, too, back when they were small. They are not so small anymore. Everything in nature has a spirit, says one of the Lenape laws, and should be given thanks, gifted, and asked permission before taking from it. In Alakanuk, Denis Sheldon, whose real name is Kituuralria, told me the same.

I doubt Jacobus Van Cortlandt, landowner, enslaver, and mayor of New York, asked permission when he had Black people he owned dam up Tibbetts Brook in 1699 to install a sawmill and gristmill on his plantation. Some of the skeletons of those he enslaved were unearthed by construction workers laying down railroad track in the 1870s. The mill operated until 1889, when the city purchased the land for its park. At that point, the millpond became a small, decorative lake. Sometimes I walk the kids to this lake, next to the African burial ground, to watch the damselflies and dragonflies hover above it, the bullfrogs breathing at its edges, and to contemplate what lies beneath.

The burial ground was consecrated with a new sign and the lake named after the enslaved miller who milled grain there, along with his wife, in a Juneteenth ceremony the summer we moved in: Hester and Piero's Millpond. A small reparation. At the south end of the lake, in 1912, the brook was piped into a storm drain and rechanneled into an underground tunnel, merging into a brick sewer below Broadway, now a main sewer line for much of the Bronx. This made it possible to construct streets and buildings south of the park, including our house, on top of backfilled dirt, probably excavated from the construction site of a nearby reservoir, possibly supplemented by the dig for Grand Central Station, and city trash. This is the ground we live on. Steve says the history of many of the city's major sewer lines can be similarly traced back to the streams and rivers that ran for centuries and even millennia before the city mushroomed around them. What does it mean to live in a place where rivers are harnessed to carry our waste away, so we don't have to think about it; to make our shit disappear?

According to the DEP, four to five million gallons of water flow into the Broadway sewer on a dry day from Tibbetts Brook and Hester and Piero's Millpond alone. It runs through the sewer, where it mixes with raw household sewage, including the shit of my family, traveling to Wards Island Wastewater Treatment Plant. But when it rains, the amount of water can be five times that. At least sixty times a year, the treatment plant gets overwhelmed by rainwater and shuts down. Untreated sewage and rainwater are then discharged into the Harlem River, in violation of federal law, contributing to river pollution.

Now, there are plans to "daylight" the subterranean stretch of Tibbetts Brook, bringing it back to the surface. This restoration will alleviate flooding by rerouting the buried section of brook directly in the Harlem River, not exactly along its historic route upon which our house sits, but slightly to the east, along an old railway line that accidentally reverted to an urban wetland after the freight trains stopped running in the 1980s—a gully running behind BJ's Wholesale Club and the strip mall with the nail salon and the Flame hibachi and the Staples, and the Target closer to 225th where we buy the boys' public-school uniforms—already rewilding with tall marsh grasses and reeds. This climate mitigation will remove billions of gallons of water from the overtaxed sewer system, an attempt at marking an end to centuries of industrial pollution and neglect.

There is talk of "undoing the past," of "resilience," of "giving some of what was taken from nature back to nature." There is talk of a bike path along a greenway costing many millions of dollars. When the plan comes to pass, it will be New York City's first daylighting story, and we will be in the watershed. It seems like a good thing. But my neighbor Alma, who was born in the neighborhood after her parents immigrated here from the Dominican Republic in the 1960s, and still lives in the house she grew up in, worries about green gentrification. She wonders, quite sensibly, who will be displaced as we remake the landscape to bring the brook back? She is not proprietary about the land. She believes we're all just guests.

As for our makeover of the house, it took over a year after closing for it to be move-in ready. We learned in that violent, hopeful year of knocking down walls and moving doorways that the pipes were leaking. That the furnace was faulty. That nobody in reach yet knew how to heat the house without hooking it up to the fossil fuel we were meant to be phasing out for the health of the planet. That my father was dying. And that the pond was there to stay. I never once thought, until talking to the elders in Alakanuk, that I should ask the pond's permission to dwell alongside it.

We were still living out of boxes when the National Weather Service declared New York City's first flash-flood emergency. The boys were by then eight and ten. Over three inches of rain fell in just one hour, shattering a record set by a storm the week before. Was it even correct to call it a five-hundred-year rainfall event when the past had become such a poor guide to the present? The remnants of Hurricane Ida turned the nearby Major Deegan Expressway back into a river, stranding cars, buses, and trucks in high water. That image, from our new neighborhood, became an international symbol of the city's unpreparedness. A two-year-old child was among the city's casualties who drowned in basement apartments. Our basement got wet, but nothing like that—not yet.

Every single subway line in the city was stalled: the 1, 2, 3; the 4, 5, 6; the 7; the A, C, E; the B, D, F, M; the G; the J, Z; the L; the N, Q, R, and W—all the letters and numbers in the grammar and mathematics of our usual movement. Some folks compared the footage of flooding in certain subway stations to Niagara Falls. A thousand straphangers were evacuated from seventeen stuck trains. "We are BEYOND not ready for climate change," declared a city council representative on social media.

The pond in front of our house was as wide and worrisome as I'd ever seen it, whipped into waves by the wind. It was as sure a sign as any that we were living on borrowed time. But in the weeks that followed Ida, against our better judgment and lacking other available options, we had Con Edison connect us to the gas line under the kettle in the street where the water gathers. We weren't yet free to make carbon-free choices. At times, the house feels like a death trap. I mean to say, if I remain embarrassed as a homeowner, it is not on account of the pond.

Yet I'd be remiss, as a homemaker, if I didn't also admit that our house feels good. I've been told it has "good vibes." Having committed to fixing the house, my husband and I salvaged and repurposed what we could. We studied magazines on home decor. We remade the kitchen as its heart. We insulated the attic. For protection, we drove an iron railroad spike into the earth at one of its four corners. We ordered light fixtures. We got housewarming gifts from new neighbors: A potted plant. A gift certificate to the Tibbett Diner. We got a compost bin from the city. We got free stuff from the local Buy Nothing group, including a mid-century modern end table and a pair of cream-colored lamps. I found a double-basin cast-iron sink on the curb outside the Ortiz Funeral Home that I sensed was a gift from my late colleague, Rosaymi, whose memorial was held there. We got a rosewood cabinet from the nursing home of Drema's dad after he passed from COVID; we brought back its luster with lemon oil. We bought two kilim rugs from a sweetheart named Hamza who sold them to us from the back of his truck. I got a hand-thrown pottery bowl from Tamar, made in Israel, to hold fruit. We painted the walls in the colors of our choosing: haint blue for the entry, smoldering red for the door. We screwed in smoke alarms. We installed skylights in the attic, to bring in the light, to show us the moon, and maybe, if need be, to climb through when the waters rise.

We built each of the boys a new bed. We unpacked our books. We unrolled our carpets. We hung up our curtains and clothes, our calendar and clocks. We put a punching bag in the basement and a solar array on the roof. We voted in the local election. In the altered rooms, we arranged the old furniture. The big oak desk at which my dad penned his dissertation now belonged to me. I placed the framed photos of my ancestors on the desktop among fresh-cut wildflowers, as a makeshift ofrenda, to ask for their blessing and guidance as I write. Watching over me: Edward Ishem, Philomena Laneaux, Nanan Patience, Mabel Sincere, Albert Raboteau, Albert Raboteau Jr.

By now, we've settled in. We threw another New Year's party, at the request of Centime, whose cancer has spread to her liver, rendering short her remaining days. We cook for guests. We lay the dining room table and light the candles. Before we eat, we pray. In the bathroom, the children have lately learned, thank God, to wash their own hair. On the doorframe, with pencil, we mark their monthly growth. On the fridge, under a magnet, we list their weekly chores.

Just as remarkable as the pond out front is the garden out back. Outside, down on my knees with my hands in the soil, I gratefully weed and tend the beds in the shade of a neighbor's pear tree. I prune the St. John's wort but leave the fleabane and aster alone. I store my tools in the cedar shed I built beneath the towering pin oak run amok with squirrels. In its understory, I plant a shadblow tree, in memory of my dad.

My mother has given me a Lenten rose. It is the first thing to bloom in spring. I marvel at the shoots coming up from the bulbs planted before me by Mary, wife of Jeremiah, whose name was not on the deed but whose name was told me by our neighbor Eve, the traveling nurse who lives on the other side of the fence. Daffodils, peonies, hyacinths, and tulips. I have added yarrow, wild columbine, bee balm, butterfly weed, creeping phlox and coral bells. My body is at ease in the garden, which replenishes like a hidden spring. In the garden, my head is correct.

I acknowledge I live in Lenapehoking, the unceded territory of Lenape people, past and present, and I further acknowledge that stories of the purchase of this land were lies told to cover up systemic slaughter. I honor with gratitude the land and the generations of people who have cared for it. I recognize their survivorship through time. I name this pattern to combat erasure and remember our shared responsibility to learn to be better stewards of the land we inhabit. I want these words to be more than words; I want them to be deeds. I'm learning to grow food for our table, sensing that the truest sacrament is eating the earth's body. The seeds were a gift from Luz, who parked Langston out front, on top of the pond, to house-sit a while, so we could finally take a break from the city and remember why we love it. (Its swagger. Its rhythm.) I have planted lettuce, tomatoes, sweet peas, and beets. I collect water in a barrel under the gutter spout. The children lift the stones to peer at the pill bugs and worms. I see in my mind's eye, as I garden, that our land is a quilt, and our house is only a structure among structures among pollinating plants visited by bees.

The pond is part of the place where we live. To prevent stagnation, I sometimes stir it with a stick. Through the front windows, I watch it swell when it rains. I observe the birds who stop there to bathe: warblers, tanagers, grosbeaks, sparrows. (*And smale foweles maken melodye, / That slepen al the nyght with open ye . . .*) Some of them are endangered. A small reparation: I am teaching our children their names.

ACKNOWLEDGMENTS

I wish to thank the following individuals, ancestors, organizations, and institutions for their support with the visioning, scaffolding, writing, research, editing, production, and promotion of this book: Retha Powers, my editor. Amy Williams, my agent. The people I interviewed who trusted me with their points of view. The street artists whose work I photographed. The neighbors whose bodies I photographed. The editors who commissioned, amplified, or otherwise stewarded prior iterations of these essays: Jesmyn Ward, Roxane Gay, Ayelet Waldman and Michael Chabon, Sari Botton, David Wallace-Wells, Leigh Newman, Morgan Jerkins, Chip Blake, Rebecca Solnit, Joshua Jelly-Schapiro, Garnette Cadogan, Sumanth Prabhaker, Lucy McKeon, Natalie Eve Garrett, Rachel Arons, Lizzy Ratner, Amy Brady, Tajja Isen, Robert Macfarlane, Ed Yong, and Saeed Jones. Members of our climate reading group: Meehan Crist, Meera Subramanian, Lacy Johnson, Bathsheba Demuth, Elizabeth Rush, and Roy Scranton. The women uptown who held me during the pandemic: Angie Cruz, Elleni "Centime" Zeleke, and Ayana Mathis. Fellow New Yorkers who've inspired and guided me: Mikael Awake and Lissette J. Norman. Climate artist Justin Guariglia. Writers Rebel NYC participants, including: Jenny Offill, Amitav Ghosh, Elliott Holt, Pitchaya Sudbanthad, Rob Spillman, and Joel Whitney. The Council on the Uncertain Human Future journalists and writers group, helmed by Sarah Buie and Kathleen Dean Moore. Director of the Climate Museum Miranda Massie. Comet-ME. Breaking the Silence. Peoples Justice for Community Control and Accountability. WE ACT for Environmental Justice. My work coach, Rena Seltzer of Leader Academic. My therapist, Alicia Fleming. My colleagues at City College, especially Salar Abdoh, Michelle Valladares, Vanessa Valdés, and Maria Tzortziou. My Climate

Writing students. Research mensches, Brent Hayes Edwards and Michael Vazquez. Copyeditor, Molly Lindley Pisani. Proofreader, Sarah Bowen. Designer, Meryl Sussman Levavi. Publicists: Sarah Jean Grimm and Caitlin Mulrooney-Lyski. Editorial assistants: Leela Gabo and Natalia Ruiz. Marketing duo: Arriel Vinson and Sonja Flancher. My graduate student assistants: Nadia Bovey and India Choquette. My photo assistant, Cristina Ferrigno. My sisters from the inaugural Freedom Writers Retreat: Dolen Perkins-Valdez, Tiphanie Yanique, Natalie Baszile, Faith Adiele, Jacinda Townsend, Lauren Francis-Sharma, and Shay Youngblood. Supporters of Black self-determination: Werten Bellamy and Kellye Walker. Sponsors: Stuart Z. Katz, the New York Foundation for the Arts, the Bronx Council on the Arts, the PSC-CUNY Research Award, the CUNY Book Completion Award, the CUNY Interdisciplinary Research Grant, Yaddo, the Sustainable Arts Foundation, and the Robert B. Silvers Foundation. The Library of Congress. Louis Armstrong and Lil Hardin. My grandmother: Mabel Sincere Raboteau for making a way. My parents: Katherine Murtaugh and Albert J. Raboteau, for making another way. Above all, the family of my making: Victor, Geronimo, and Ben. We're making our way.

At the Guggenheim

ABOUT THE AUTHOR

Emily Raboteau writes at the intersection of social and environmental justice, race, climate change, and parenthood. Her previous books are *Searching for Zion*, winner of an American Book Award and a finalist for the Hurston/Wright Legacy Award, and the novel *The Professor's Daughter*. Raboteau is a contributing editor at *Orion* and a regular contributor to the *New York Review of Books*. Her essays have appeared in the *New Yorker*, the *New York Times*, *New York*, the *Nation*, and elsewhere. She is a professor of creative writing in the English Department at the City College of New York (CUNY) in Harlem and lives with her family in the Bronx.